NEW WINE
REVELATIONS

New Wine Revelations

Inspired and Dictated by
God Jehovah, Exodus 6:3 - Creator
Author
Architect
Savior
Teacher
Healer
Divine
Eternal
Friend

George W. Chivers
Disciple, Secretary, Friend

Rev. date: 09/20/2016

To order additional copies of this book, contact:
Xlibris
1-888-795-4274
www.Xlibris.com
Orders@Xlibris.com
715600

Greetings to Friends Worldwide

Hello	-	English
Aloha	-	Hawaiian
Yasoo	-	Greek
Shalom	-	Hebrew
Hola	-	Spanish
Asalamalakem	-	Arabic
Konnichi wa	-	Japanese
Chow	-	Vietnamese
Ciao	-	Italian
Salundt	-	French
Hafa day	-	Guamanian
Hallo	-	German

Contents

Total: (1) 1000 years + 2000 years + 500 years = 3,500 years

(2) 720,000,000 years + 1,440,000,000 years +
360,000,000 years = 2,520,000,000 billion years

- Converting Old Testament (OT) patriarch coded ages into decoded ages: example, Adam's coded age was 930 years:

930 years x 360 = 334,800 years (minor calculation)
930 years x 360 x 2 = 669,600 years (medium calculation)
930 years x 360 x 2 x1000 = 669,000,000 million years (major calculation)

Note: We're living in God's rest day which may also be 720,000,000 million years. We have only lived in God's rest day which is 236,880,000 million years. Therefore, Adam may have lived only in the minor or medium age range, not major.

- Syllabic analysis of prefixes, infixes, and suffixes in the name Eloheem

- Earth was created out of water as stated by Moses in Genesis 1:2, Apostle Peter in 2 Peter 2:5, and geophysicists.

- Moses predicted that Israel would discover oil and gas in the later days while blessing the children of Israel, that one tribe would dip their feet in oil and wear iron shoes (chemical engineers in the present).

- A cocoon of water once hovered above and around the earth and firmament; every square foot of the earth was tropical, and the planet was *slightly* closer to the sun. The gravitational pull of the sun and earth kept the earth in balance.

- Copy of e-mail that was sent to epidemiologists at CDC NIH with suggestions about Ebola virus oxidized methemoglobinemia

- Dr. Madison Cawein treated blue skinned residents of Kentucky with E. Methylene Blue which reduced the oxidized +3 iron back to +2 ferrous irons.

- Eosin Methylene Blue was used as an antimalarial drug, may be used as an antiparasitic emetic drug; medical trials should be conducted on Ebola infected patients. Ebola, like female malarial mosquitoes that have a desire for host blood and live in blood of vampire bats, may have a female character and are fertilized by fruit bat male Ebola viruses. God made male and female species. (*during World War 2)

- Excessive blue light has been found to be harmful to the retinas of mice, humans, and other animals. Painting porch ceilings blue keeps mosquitoes away.

- The color of Adam and Eve was ruddy and rosy; Cytochrome C Oxidase

- Melanin and different hues in humans

- Noah's Ark, family, and conflicts; Mormon Prophet Lehi traced through Jacob, Joseph, and Manasseh to Syrian relatives who wrote and spoke Aramaic. Aramaic may have been the sole language before the Tower of Babel.

- America and Britain in prophecy. Manasseh is America and Ephraim is Britain; ref, book written by Pastor Herbert Armstrong, Sr., founder of the Universal Church of God, Pasadena, CA

- Tracing genealogy of Mormon Prophet Lehi through Jacob, Joseph, and Manasseh in Old Testament (OT), Strong's Concordance, and book of Mormon

- Early Mormon Church struggled against clannishness to admit people of Semitic and Hamitic races.

- Present Church of Jesus Christ of Latter Day Saints (CJCLDS) sends missionaries of all races, two by two, to all countries around the world.

- Polygamy banned in Mormon Church

- Noah's Ark and covenant with God to build that unique flood-faring structure

- Noah's children and their descendants. Hamitic people made servants of Shem and Japheth. Hamitic people denied the role of priesthood because of an offense toward Noah and his wife until the advent of Jesus

- Shem may have been Melchizedek and lived thousands of years beyond the coded age of five hundred years; his, Japheth's, and Ham's descendants, who had been decreed by God to live only a maximum of 120 years, might have thought Shem was godlike—without birth, without parents.

- Anopheles and Plasmodium malarial diseases

- Hosts' diploids and mosquitoes' diploids combined and calculated using a model to obtain macrogametes, and microgametes using the mitosis telophase 1 and the meiosis telophases 1 and 2

- Crustaceans chromosomes diploidy, mitosis telophase 1, and meiosis telophases 1 and 2 calculations

- The single fertilization of the female malarial mosquitoes by the fruit-eating male mosquitoes in a swarm in the air

- Plasmodium falciparum 14 chromosome diploidy. Anopheles mosquitoes have a six diploid, three haploid.

- (Redox) Oxidation and reduction reactions involved in the control of the malarial female vectors and the treatment of the malarial disorder

- Minus (-) 3 magnesium ions needed to reduce +3 ferric iron back to plus +2 ferrous iron in the malarial cycle

- Malarial ovum/protozoa cycle and yeast growth cycle similarity

- Female malaria mosquito and fruit-eating male mosquito single fertilization encounter

- Abortive effects of excessive quinine and excessive caffeine in vertebrates

- Cytochrome P-450 system a.k.a. CYPP-450 can be activated and deactivated by either inducers or inhibitors (Lavinol in grapefruit is an inducer a.k.a. enhancer); Taxanes slow down mitosis; star fruit and watercress are inhibitors; cigarette smoke and ethanol are very bad inducers.

- Star fruit and watercress are inhibitors which contain 1,000 to 4,900 international units of vitamin A.

- The Deluge and Noah's Ark continued from page 13; the floodwater rose above the mountains and all creatures not in the Ark drowned, no aquatic creatures were taken aboard the Ark.

- If Alphaeus, a.k.a. Clopas, were the one and only Alphaeus in the New Testament, then he could have been the stepfather of Jesus and the father-in-law of Apostle Matthew; fathers-in-law could claim their sons-in-law as sons, and grandfathers could claim their grandsons as sons.

-The Jewish Siva: 3 days + 7 days + 30 days = 40 days

- What was the gender of Apostle John, a.k.a. the apostle that Jesus loved, also referred to as the Apostle of Love

- In the Second Epistle of John, He wrote about the "Elect Lady" and "Your Elected Sister." Was he referring to himself or the Church?

- Where the apostles were seated around Jesus at the Passover table especially John, James, Peter, and Judas Iscariot?

- John wrote about the female disciples who followed Jesus to the cross. John did not mention that he was among them. Was he dressed as a woman? He was the only apostle at the cross.

- The Bible says that the apostle that Jesus loved knew the high priest. No one questioned "John" like they did Peter. Did John blend well with the women present?

- Joanna stood near the cross with the other female disciples. Jesus addressed His mother as "woman" and Jesus addressed "John" as "son." Did Jesus declare John's gender status as a "male?"

- Surname(s) of Apostle Jude and stepbrother of Jesus, Labbaeus in Hebrew and Thaddaeus in Greek. Therefore, Jesus's whole family would have had these surnames.

- The name John means Jehovah/Yahweh is gracious, and the name Joanna also means Jehovah/Yahweh is gracious.

- In Psalm 22, prophet and poet David prophesied Jesus's crucifixion and resurrection eight hundred years before His birth. Jesus's ten apostles, excluding Judas, abandoned Jesus, not God. The ten apostles hid because they were afraid of the Jews and the Roman soldiers.

- Zephaniah 1:15 and Zechariah 14:1, 6, 7 prophesied that on the great day of the Lord, there would be gloomy darkness in the day and light in the evening.

- God shortened time during Jesus's thirty-six hours entombment, three twelve-hour days. It appears that God extended time during Joseph of Arimathea and Nicodemus's burial preparation of Jesus's *dead* body. John 19:40 NKJV Sundown may have come around 9 PM instead of around 5 PM.

- Rabbinical teachings in those days were that Heaven was above in the sky and that paradise was down insheol (hell) a.k.a. Abraham's bosom divided by an unbreachable chasm. The *three* went to paradise.

-The "other" Mary, mother of Jose, wife of Cleopas a.k.a. Alphaeus's sons were James, Jose, Jude, and perhaps Simon. Alphaeus and Simon walked seven miles from Jerusalem to Emmaus when they met the risen Christ. That Simon may have been Cleopas's son. When they returned to Jerusalem, Cleopas said Simon *believed* in Jesus and His resurrection. Mark 15:47 NKJV the syntax is misleading. This Mary may be Jesus's mother.

- Church historians claim that Mother Mary was in Gaul and Lourdes, France and has appeared there and to countless people around the world.

- Pontus Pilate regretted Jesus's death and believed Jesus was a "just person" and wrote He was "King of the Jews" which automatically said He was the Messiah.

- Jesus may have been nailed to a single crossbeam on the ground and hoisted up to a stationary and permanent cross with a fixed crossbeam, in less than ten seconds. Wood was scarce in Jerusalem.

- The thirty-six hours crucifixion timeline and possible visible weather conditions (VWC) in Jerusalem and around the planet

- Origin of the holy name *Jesus*; there are no "Js" in the Hebrew and Greek alphabets. In Greek, Jesus's name is Iesous and Yeshua in Hebrew.

- Origin of the holy name Allah. The Arabic language was derived from Aramaic. Alaha which means God in Aramaic was Arabized using the shadda (W) over the (L) and removing the final (A) which is the article (the).

- September was the original birthday month of Jesus Christ. He was thirty-three and a half years old on Nisan 14th. When we count backward to six months, He was thirty in September. The Church claim that December ninth was the Inception Day of Jesus; count forward to nine months and we come to around September ninth

- Book highlights and commentaries

- Gratitude and Epilogue

Introduction

When a spectator asked Lord Jesus Christ, "Of all of the laws of God, which is the best law?"

Jesus responded, "You shall love your God with all your heart,* all your soul, and with all your mind" (Matthew 22:37 KJV). Jesus continued, "This is the first and great commandment" (Matt. 22:38 KJV). "And the second is like it: you shall love your neighbor as (you love) yourself" (Matt. 22:39 KJV). "On these two commandments, hang all the laws and the prophets. This is my commandment, that you love one another as I have loved you." *with . . .

Note: Our neighbor is everyone on this earth.

Divinity of Jesus Christ as cocreator with God

John 3:16 KJV – For God so loved the world that He gave His only begotten Son, that whoever believes in Him should not perish but have everlasting life (for eternity).

John 1:1 KJV – In the beginning was the Word, and the Word was with God, and the Word Was God.

John 1:2 KJV – He was in the beginning with God.

John 1:3 KJV – All things were made through Him, and without Him, nothing was made that was made.

John 1:4 KJV – In Him was life and the life was the light of men.

John 1:5 KJV – And the light shines in the darkness and the darkness did not comprehend it.

John 1:10 KJV – He was in the world and the world was made through Him and the world did not know Him.

Isaiah 9:6 KJV – For unto us, a child is born, unto us a son is given, and the government will be upon His shoulders. And His name will be called Wonderful, Counselor, *Mighty God*, *Everlasting Father*, and Prince of Peace.

Autobiography Synopsis

Hello,

My name is George Waymond "Wayne" Chivers II. I was born on January 01, 1939, at 3:00 PM in Carrollton, Georgia. Today it has become a farming, industrial, and logging city, fifty miles southwest of Atlanta, Georgia.

My mother, Laura Gibson Boone, her mother, Fanny Mae Merdis, and my father's mother, Mary, were born with the gifts of clairvoyance, remote viewing, and remote hearing a.k.a. clairaudience. I have also been blessed with these gifts and talents. These gifts are most effective in all people when our spirits have been quickened and uplifted by prayer, music, conversation, exercise, and love, and when we are in an alpha wave state of mind and grace.

My genealogy consist of Western and Middle African, Welsh (Phillips), Scot-Irish (Gibson/Johnson), English (Chivers), Seminole, Greek, Cherokee, French (Chevre), and Italian (Capa). My interests are Christ unification of all religions, all research, health and environment, education, and welfare worldwide.

My hobbies are sudoku, pinochle, writing clean jokes, ballroom, and Latin dancing.

Sincerely,
George W. Chivers II

The One and Only Way
to God the Father

Thomas said to Him, "Lord, we do not know where you are going, and how can we know *the way?*" (John 14:5 NKJV).

Jesus said to him, "I am the way, the truth, and the life. No one comes to the father except through me" (John 14:6).

In our present earthly form and state of mind, we cannot be in the presence of God the Father. God's light and glory would consume us. In the book of Revelation, Jesus told Apostle John that when we are transformed into our immortal bodies, God will live with the just believers on the new earth.

Presently, Jesus is our spiritual transformer. An electrical transformer can power up voltage and reduce voltage, so weaker appliances can use a much lower voltage and not burn out and be destroyed. Jesus has never lied and never will. Jesus is the Son of God and has godlike power and authority. "The truth shall set you free." The scriptures confirm that God the Father and God the Son cannot lie:

> Thus God, determining to show more abundantly to the heirs of promises the immutability of His counsel, confirmed it by *an oath.* (Hebrews 6:17–18 NKJV)

> That by two immutable things, *in which it is impossible for God to lie*, we might have strong consolation who have fled for refuge to lay hold of the hope set before us. (Hebrews 6:18)

If we consider Father God as a lake of pure spring water and a portion of the spring water is drawn out, then this portion of the pure water would represent Jesus the Christ. Thus, Jesus would have a composition of 100 percent as pure as Father God. Then this analysis would also make Jesus God like the Father. Therefore, when we want our prayers and requests to reach God the Father, we have to send them, first, to Jesus Christ the Son of God. Amen

Disclaimer

We disclaim any copyrighted publications that have been used in *New Wine Revelations* that we failed to reference due to an oversight on our part. You, researchers and authors, know your copyrighted materials, and we insist that you have the lawful right to claim any part of it at any time. Any material that we have submitted and claim as original that you have researched and copyrighted, we disclaim as being original. As *New Wine Revelations* goes to print, we will endeavor to obtain permission and list all numerical references that we have used in our book NWR. We apologize for any copyrightable infringement we have caused.

Sincerely,
NWR Authors

Sincere Apology to the Readers

We apologize for any inconsistencies, redundancies, and typographical errors that you might encounter during the reading of this book. A thorough reediting in order to eliminate these literary mishaps, we decided to set the publication of NWR back several months. This book was written, at times, under dire conditions with many distractions and interruptions. To justify this trendy literary action, we recalled the story of a good citizen who was walking down a street and saw a house on fire. Thinking he/she was not a good orator, this person went back to school for four years to study public speaking. This person graduated at the top of his/her class. Then this person returned to the street where they saw the house on fire to proclaim in an oratorical manner that the house was on fire. Unfortunately, the house had burned down. The space where the burning house once stood was now a parking lot. The moral to the story is it's not how important the way you say something, but it's important to say it right away. Please treat NWR Book like the burning house. Thank you.

Dear Friends,

Greetings in the names of our Father God Yahweh, Lord Jesus Christ, and the Holy Spirit also known as the Counselor, Advocate, and Comforter

During the eons of eternal time, Father God Yahweh created, in this order, the heavens, universe, earth, moon, stars, and all stellar systems according to Prophet Moses in the Judeo-Christian book of Genesis. KJV

The Bible says there was light in the universe before the stars were created. Astrophysicists and astronomers also claim there was illumination in the heavens before the stars were created. This is one of many cases where science and religion agree. Astrophysicists and geophysicists claim that the universe is about 13.5 billion years old, and the earth is close to 4.5 billion years old. They base their theories on the movement of the celestial system, meteorites falling to earth, and radioactive rock decay. Religion also confirms the ages of the universe and earth that science has reported and documented in various journals, books, and periodicals. The religious equation for decoding prophetic time into scientific time is as follows:

Coded prophetic time in years x 360 solar years x 2 = days x a day with the Lord = astrophysical decoded time in years. One decodified creative day in God's calendar may equal to 720,000,000 million years.

The coded biblical year for 2016 is 6329. When we divide 6329 by pi (3.14), we get the Gregorian year 2016. The year 6329 indicates that we are now living in God's rest day.

Syllabic analysis of the prefixes, infixes, and suffixes in the Divine God word Eloheem:

El	=	God in Hebrew
Elo	=	God in Aramaic (Eloi means my God in Aramaic)
Hee	=	She in Hebrew
Heem	=	They in Hebrew
Em	=	Mother in Hebrew (Emma means Mommy in Hebrew)

Syllabic analysis of the prefixes, infixes, and suffixes in the Divine God word Allahum:

Al   =   In Arabic, it means "family" or "the house of. . ."
Llah =   God in Aramaic, Farsi, and Arabic
Hu   =   He in Hebrew (Hu-wa means he in Arabic)
Hum =   They in Arabic
Um   =   Mother in Arabic (Umma means Mommy in Arabic)

The book of Genesis says in the beginning that God created the heavens and the earth. The earth was without form and void and darkness was on the face of the deep. And then the Spirit of God was hovering over face of the waters. It appears that the earth in its initial stage was completely water. Many geophysicists claim that at one time the earth was predominantly water. Again, science and religion agree. In 2 Peter 3:5, Peter says that the earth was standing out of water. In the desert, God commanded Moses to speak to the stone for water. Instead, Moses hit the stone with his staff. Water came out, but God was displeased with Moses because he did not speak to the stone as God commanded him. For this infraction, and others, and because Moses had reached the prophesied age of 120 years, he was not permitted to enter the Promised Land. On a mountain, God allowed Moses to look across the Jordan River and see it. While Moses was on the mountain saying his farewells to the children of Israel, God revealed to Moses that in the lateral days, Israel would discover crude oil and gas near Mount Carmel and in the Mediterranean Sea. Reference was made to a tribe that would wear iron shoes and dip their feet in oil. Chemical engineers and oil field workers wear steel toe shoes.

Based on this oil prophecy, modern day Israel has been finding crude oil and gas. It is estimated that Israel may have as much or more per country as its neighboring countries. This information can be found on the internet under the title, "Oil discovery in Israel."

In the book of Genesis, God created the firmament and commanded that there would be water below the sky and water above the sky. The water above the firmament resembles a cocoon of water. That means that the earth must have been slightly closer to the sun so that the gravitational pull of the sun would hold the earth's water cocoon suspended with an upward force, and the earth would have a downward force effect on the cocoon. Therefore, every square foot of the earth might have been mildly tropical. That may be why we can find fossils and crude oil all over the earth.

With the earth being closer to the sun, the side of the cocoon between the sun and the earth had to be very hot. In order to cool the cocoon, the cocoon above the earth must have spun in an elliptical direction into the coldness of space and back again between the earth and the sun. When the water for the deluge fell from the sky, the the water covered the whole earth. Moses wrote that the flood waters rose above the mountains and every living creature that flew in the sky and moved upon the earth that was not inside the ark drowned.

Subject: Methemoglobinemia

Etiology: Oxidized ferric +3 iron in methemoglobin initiated by oxidized ferrous +2 iron caused by chemicals, bacterial, and viral infections

To: CDC and NIH Epidemiologists

Re: E-mails sent

Date: Saturday, 10/17/2014

Dear Epidemiologists,

With the supreme help of God Almighty, extensive online research and viral profiling, it appears that the present strain of West African Ebola virus is possibly an anaerobic retrovirus. In vivo, the virus oxidizes the active oxygen (O2) carrying ferrous +2 iron in normal and abnormal hemoglobins (HBa, HBs, HBc, HBf, HBm) to the inactive non-carrying O2 ferric + 3 iron. The ferric +3 ions form methemoglobin in human blood. In concentrations greater than 15% dc/l, the methemoglobin forms a blood disorder called Methemoglobinemia

Subject: Methemoglobinemia

A medical doctor who treated the actual "blue skinned people" of Kentucky (re: Martin Fugate and relatives) discovered he could reverse the inherited ferric +3 iron blue skin color by injecting his patients with an isotonic 1% eosin methylene blue titration indicator dye. First, he and other doctors tested for allergic reactions by performing a skin test with a bandage application, and by injecting a small bubble of the 1%

solution under the skin and looking for swelling and/or discoloration in twenty-four hours. Also, the attending doctor and medical colleagues have warned that 1% methylene blue solution therapy should not be administered to patients with low levels of glucose-6-phosphate dehydrogenase (G6PD). People with low G6PD show hemolysis of red cells among other side effects. 1% methylene blue by IV can turn inherited and acquired blue skins back to pink in fair skinned people and a lighter hue in darker skinned people.

The IV method can turned ferric +3 iron ions back to ferrous +2 in a matter of three hours. Orally, in a matter of few days. The subject doctor injected five cubic centimeters (CC) over a period of one hour.

Subject: Methemoglobinemia

Fruit bats eat only nuts and overripe oxidized fruits. Fruits contain, among other chemicals and nutrients, iron and magnesium. The oxidation/reduction potential of Fe +3/Fe +2 = a positive + 0.77. Mg +2/Mg = neg - 2.36 Mv potential. Pos + 0.77 added to Mg neg -2.36 =-1.59 Mv potential. Because the millivolt potential is neg, Fruit Bats don't acquire Ferric + 3 iron methemoglobin.

In addition, halide salts ions such as -Cl, -F, -Br, -I have negative ox/red potentials. In vivo, these ions prevent retroviruses from converting their RNA uracil to DNA thymine. The halide ions prevent the attachment of a negative charged methide from attaching at the fifth carbon site of the uracil. The ions also prevent the dihydroxylation removal of one alcoholic —OH group from the uracil due to an increase in acidity by ferric +3 irons.

Optimal electron reduction and electron reductases are needed when reducing excessive oxidized products in vivo and vitro.

Subject: Methemoglobinemia

It has been demonstrated in high school and college labs that electrons will flow through a group of students holding hands and hooked to a storage battery. A millivolt meter attached in the line will show that a flow of electrons is occurring when the volt meter's needle flickers.

Theoretically, beta radiation, which has electron properties, should be able to reduce highly oxidized substances and materials by flooding them with beta rays. Also, beta radiation should be able to treat abnormally acidified health conditions, such as cancer, bacterial and viral infections. Electron beta radiation should promote cancer cell apoptosis and keep bodily reactions moving in their normal forward direction that is from the left with reactants to the right where products form. Thank you very much for reading these suggestions. I pray that they will be useful in future trials and research.

Sincerely,
Verizon.net
George W. Chivers, R&D Medical and Chemical Engineering Technologist

Adverse Effects of Blue Light to the Eyes of Mice and Humans

Medical researchers have found that Methylene Blue is beneficial for the eyes of mice and humans. However, they have found that excessive blue light is harmful to both mice and humans; maybe other mammals,. A team of eye surgeons at Heriot-Watt University and Edinburgh's Princess Alexandra Eye Pavilion claim that blue light inhibits the Cytochrome C Oxidase in eye cells. This enzyme is the last protein complex that helps in the respiration of the eye cells and permits a transfer of electorns across a transmembrane called the mitochrondrion. In this "little organ" is where Adenosine Triphosphates (ATPs) are made. Every chemical reaction and movement in animals require ATPs.

These eye surgeons claim that blue and green eye people have less Macular Pigment than other colored eyes. Blue eye retinas act as blue filters, allowing blue light to pass through to the optic nerves. After the age of 40, Age-related Macular Degeneration (AMD) begins to show during eye examinations.

Many medical doctors, who use ophthalmoscopes for eye exams, insert yellow filters inside the scopes which prevent blue light from being projected into patient's eyes.

Many ophthalmologists prescribe yellow tinted prescription glasses and sun glasses to protect their patients eyes from excessive blue light and UV light exposure. There is a lot of natural and artificial light around us. The sky at times and some oceans are blue. High tech game machines, computer screens and colored televisions project a lot of blue light. The U.S. Naval personnel, who are posted on ships and have to stand a watch topside, are required to spend fifteen minutes in a low illuminated infrared room right before assuming duty. This prewatch procedure enhances the eyesight in the dark and counteracts the effects of the blue sky.

Many nutritionists claim that kale, spinach and cabbage contain sufficient amounts of "macular pigment." Scientific studies show that a room which has been flooded with blue light will deter mice and rats from entering that area at night or when that space has been closed up and rended light free. The blue light bulbs MUST be attached to electrical sensor units so that when someone turns on a luminescent (white) light in the room that contains blue light, the regular light will make the sensors shut off and prevent the blue light bulbs from emitting blue light. This precaution is done so that when people enter a blue light room they will not subject their eyes to prolonged blue light exposure. Then, when someone turns off the regular lights and leaves the room, in a fraction of a second, the blue light bulbs will turn on. Just imagine, using blue light, whichappears on the color light spectrum at 450 nanometers (nm), to rid rodent pests from infected enclosed areas and outside dark areas; that is, no more poison pellets and unsightly traps in your home, business and outside dark areas. Researchers claim that most animals are color blind. Are other animals' eyes, including insects, affected by blue light? Some people claim that when they paint their porch ceilings with certain blue paint, it keeps mosquitoes away.

Primary Colors for Pigments and Light

The primary colors for pigments are red, blue, and yellow. The primary colors for light are green, violet, and orange. These colors are found in the spectrum. Also, all six colors are found in the rainbow. Neither white nor black are found in these light mediums. Artists and spectroscopic analysts claim that white and black are "noncolors." When artists want to make black, they mix equal amounts of, first, red, then blue and yellow together in a nonporous container. What was the color of the first humans that God and Jesus Christ created?

Adam does not mean man. Ish is the name of man in Hebrew and ishbar for woman. Adam means ruddy, red like the earth. All of the first parents and their offsprings had rosy skin. Then when Cain slew his brother Abel because the Creator favored Abel's animal offering over Cain's grain offering, God banished Cain to Nod, the place of wandering. Cain cried out to God, "Whomever I will meet, they will kill me." Biblical scholars believe that God gave Cain the rare complexion of black so that people would see Cain as unique and want to protect him and not destroy him. A black complexion is an aesthetic condition of beauty.

Case in point, recall the beauty of a black orchid, a black rose, and a black tropical fish swimming among other different colored tropical fish, darting in and out of columns of multicolored coral over beds of natural painted sea shells. This is a picture of God's beauty in gardens and in the sea. What makes the skin of tropical fish, mammals, flowers, and even some bacteria black?

There is a test in the bacteriology lab called the Cytochrome C Oxidase Test. Under sterile conditions, a sample of bacteria is smeared on a strip of filter paper that has been impregnated with cytochrome called an oxidase reagent. A reduction reaction takes effect by the emission of electrons from the bacteria to the cytochrome. Within twenty-five seconds, the impregnated oxidase reagent paper, if positive for the presence of oxidase, will turn violet and dark blue. When no oxidase is present, the paper will not change color. If the positive test is allowed to remain after twenty-five seconds, the positive test paper will turn black. This is an in vitro test. An in vivo oxidase detection in mammals would most likely take place at birth or before. A black complexion would remain for the life of specific species. Further studies should be done to confirm this complexion theory.

Melanin, which is produced by the oxidation of the amino acid tyrosine, is another substance that can produce darker hues in humans. There are three types of melanin: eumelanin, pheomelanin, and neutromelanin. Eumelanin is responsible for the pigmentation in people with black and brown hair. Pheomelanin is a cysteine amino acid oxidized melanin that is found in people with reddish and blond hair. People with pheomelanin have freckles on their faces, chest, and upper back regions. Neutromelanin is associated with "gray matter" of the brain. Other than that, little is known about neutromelanin. Melanin of all types protects the skin and eyes from harsh sunlight rays. Albinism syndrome is a genetic condition where melanin is absent in a person's skin, eyes, and hair. The "gift of melanin" is another protection that the Creator(s) gave Cain and his descendants. God knew, and planned, that during His creative seven-day week, the water cocoon circulating His earth would come "crashing" from the sky in the form of rain thus creating the flood.

Noah's son Ham, who was a Sethite, married a Cainite. Her Name was Egyptus. She carried the eumelanin gene in her genome. Ham's daughter was also named Egyptus. His daughter was credited with founding the land of Egypt. Strong's Concordance defines Egypt as black. This story is found in the book of Abraham contained in the Mormon's Holy Book. The book of Mormon has its beginning in the Judeo-Christian Bible. It started with Joseph, son of Jacob, the patriarch. Joseph had two sons with his Egyptian wife. The sons were named Manasseh, the oldest, and Ephraim. When Joseph asked his father, Jacob, to bless his two sons, Jacob crossed his arms and placed his right hand on Ephraim's head and his left hand on Manasseh's head. Joseph protested. He knew Jacob's right hand should be on the elder, Manasseh. Jacob resisted. He said that Ephraim would be the founder of many nations and that Manasseh would be the founder of a great nation.

Jacob's prophecy occurred with the rise of the British Empire and the founding of the United States of America. Pastor Herbert Amstrong Sr., founder of the Universal Church of Christ in Pasadena, CA, left the world a great legacy. During his life, he researched and wrote "England and America in Prophecy" and where the apostles chosen by Lord Jesus Christ traveled as they found and taught the "Lost Tribes of Israel."

Prophet Lehi and other Mormon prophets prophesied in the golden plates, before they married the daughters of Ismael and sailed across the Great Waters, that in the later days, a seer would be born with the name Joseph. Joseph Smith Jr. is that seer, many people believe. Lehi and others wrote on the Golden Plates, "We have the Words of the Hebrews and the Writings of the Egyptians" —Ancient Aramaic. In the Judeo-Christian Bible, there are Syrians in the genealogy of Masasseh. Also, Laban, which means white in the Strong's Concordance, spoke Aramaic and worshiped idols. Laban is found both in the book of Mormon and the Old Testament of the Judeo-Christian Bible. Present day Syrians speak modern Aramaic. Lehi, which is Aramaic, can be found in the Judeo-Christian Bible by tracing the genealogy of Joseph through Manasseh to his grandson Lechiy which is Hebrew. In Strong's Concordance, reference 3895–3896, Lechiy means fleshiness, jawbone, which is translated in the Concordance as Lehi. In 1 Chronicles 7:19, we find the name Likhi, one of the three sons of Shemida, son of Manasseh. Another source says that the name Liqchiy is the transliterated word for Lehi. Ref: KJV Old Testament Hebrew Lexicon. In the Internet, biblestudytools.com says that Manasseh's four sons were Ahiam, Shechem, Likhi, and Aniam.

When Joseph Smith and his friend Oliver Cowdery were permitted by God via the "trekking" angel, they found an Urim and a Thummin with the Golden Plates of writings. The Urim and Thummin were used by the ancient Hebrew priests to divine and communicate with God. Prophet Joseph Smith also had the gift and ability to access God's Askaskic records as the modern day Prophet Edgar Cayce of Virginia Beach, VA, USA. God writes on the Askaskic records the past, present, and future events of man. In addition, Joseph Smith and his recording secretary, Oliver Cowdery, claimed that Jesus Christ visited them in a vision. Therefore, the holy words that Joseph Smith dictated and Oliver Cowdery recorded appear to have come from divine sources provided by Almighty God.

Subject: Methemoglobinemia

Fruit bats eat only nuts and overripe oxidized fruits which may contain a trace to 4% alcohol. Fruits also contain, among other chemicals and nutrients, iron and magnesium. When the +2 ferrous irons loses an electron and is oxidized to +3 ferric iron, the +2 magnesium, if present, donates an electron and reduces the +3 ferric iron back to +2 ferrous iron.

The redox potential for +3/+2 iron = a positive +0.77millivolts (mv). The oxidation/reduction (redox) potential for +2 mg/mg = a negative -2.36 mv. Positive +.77mv added to negative-2.36mv Mg = -1.59 mv. When the mv potential is negative, fruit Bats don't acquire +3 ferric irons methemoglobin.

In addition, halide salts ions such as Cl-, F-, Br-, & I-, have negative redox potentials. In vivo, these ions prevent retroviruses from converting their RNA uracil to DNA thymine. Negative charged methides are prevented from attaching at the fifth carbon site of the uracil. Also, dihydroxylation of alcoholic groups from uracil is prevented.

Optimal electron reduction and electron reductases are needed when reducing excessively oxidized products either in vivo (the body) or in vitro (the lab).

Subject: Methemoglobinemia

In school physics and chemistry labs, electrons will flow through a line of students holding hands while hooked up to a storage battery. A millivolt meter attached to the line between the last student and the positive electrode of the battery indicates that electrons are flowing by the flicker of the volt meter's needle.

Theoretically, low beta radiation, which has electron properties, may be able to reduce oxidized substances and materials by flooding them with beta rays. Also, beta radiation should be able to treat and reduce abnormally acidified health conditions, such as certain types of cancer, bacterial and viral infections. Electron beta radiation may promote cancer cell apoptosis and keep bodily reactions in a normal forward direction that is from the left where reactants are introduced and to the right where products are formed. Thank you very much for reading these suggestions. We pray that they will be useful in future clinical trials and research.

Sincerely,
George W. Chivers, R&D Medical and Chemical Engineering Technologist

Dr. Madison Cawein, the Blue Skinned Resident of Troublesome, Kentucky, and Methylene Blue Therapy

Dr. Madison Cawein was a hematologist and professor at the University of Kentucky. As a hematologist, Dr. Cawein knew about the adsorptive and absorptive staining properties and effects of most dye especially Eosin Methylene Blue (MB). Since its creation by Heinrich Caro in 1876, MB has been used to treat various medical disorders. During WW II, MB was used successfully as an antimalarial drug. It was disbanded until recently because soldiers and civilians didn't like having blue eyes, blue and green urine. A renewed interest in antimalarial therapy with chloroquine has been showing good results in clinical trials. Besides cost, antimalarial drug resistance is a problem in eradicating malaria worldwide.

Dr. Cawein set out to find and study these blue skinned people. Nurse Ruth Penograss assisted him. Low dosages of MB IV injections returned blue skin back to a rosy color in three hours. Oral dosages took few days. MB researchers have found that megadoses of MB act as an oxidizing agent. Hemoglobins go from a ferrous iron +2 to an adverse ferric iron +3. Whereas, a low dosage of MB acts as a reducing agent. With the transfer of one electron, ferric iron +3 reverts back to normal ferrous iron +2. Optimal dosage is very important in the administration of any medication.

Adverse Effects of Blue Light to the Eyes of Mice and Humans

Medical researchers have found that methylene blue is beneficial for the eyes of mice and humans. However, they have found that excessive blue light is harmful to both mice and humans; maybe other mammals also. A team of eye surgeons at Heriot-Watt University and Edinburgh's Princess Alexandra Eye Pavilion claim that blue light inhibits the Cytochrome C Oxidase in eye cells. This enzyme is the last protein complex that helps in the respiration of the eye cells and permits a transfer of electrons across a transmembrane called the mitochrondrion. In this "little organ" is where Adenosine Triphosphates (ATPs) are made. Every chemical reaction and movement in animals require ATPs.

These eye surgeons claim that blue and green eye people have less macular pigment than other colored eyes. Blue eye retinas act as blue filters, allowing blue light to pass through to the optic nerves.

The Color of Adam and Eve and Cytochrome C Oxidase

Biblical scholars believe that Lord God, in order to protect Cain, gave Cain the rare complexion of black. This way, people would see Cain as unique and want to preserve and protect him from any harm. A black complexion is an aesthetic condition of beauty. Case in point, recall the beauty of a black orchid, a black rose, and a black tropical fish swimming among other different colored tropical fish, darting in and out of columns of multicolored coral over beds of naturally painted seashells. This is a picture of God's beauty in His gardens and in His seas.

What makes the skins of tropical fish, mammals, flowers, and even some bacteria black? There is a test in the bacteriology lab called the Cytochrome C Oxidase Test. Under sterile conditions, a sample of bacteria culture is smeared on a strip of filter paper that has been impregnated with cytochrome oxidase reagent. A reduction reaction effect takes place. Electrons are transferred from the bacteria, which is oxidized by losing electrons, to the cytochrome on the impregnated paper. Within twenty-five seconds, the impregnated oxidase reagent paper, if positive for the presence of oxidase, turns violet and dark blue. When no oxidase is present, the paper will not change color. If the positive test paper is allowed to remain at room temperature for more than twenty-five seconds up to one minute or more, the paper will turn *black*. Black skin and black fur genes are carried in the DNA of animals.

Melanin and Different Hues in Humans

Eumelanin is responsible for the pigmentation in people with black and brown hair, researchers claim. Pheomelanin is a cysteine amino acid oxidized melanin that is found in people with reddish and blond hair. (Re: Wikipedia.) People with pheomelanin have freckles on their faces, chest, and upper back regions. Neutromelanin is associated with the brain's "gray matter." Other than that, little is known about neutromelanin. All types of melanin protect the skin and eyes from harsh sunlight. Albinism syndrome is a genetic condition when melanin is absent in a person's skin, eyes, and hair.

(Re: Wikipedia.) The "gift of melanin" is a protection that the Creator(s) had given to Cain, Seth, and all their descendants. God knew, and planned, when He removed the water cocoon that circled the earth, mankind would need the protection of melanin. When He commanded the earth to move away from the gravitational pull of the sun, the windows of heaven opened up and the water above the firmament came down in the form of rain thus creating the flood, also called the Deluge.

Noah's Family, Conflicts; Jacob and Joseph; Prophet Lehi and the Holy Book of Mormon

Noah's son Ham, who like his father was a Sethite, married a Cainite, a descendant of Cain. Her name was Egypt. She carried the eumelanin gene in her DNA. Ham's daughter was also named Egypt. Ham's daughter was credited with founding the nation of Egypt. Strong's Concordance defines Egypt as black. This story is found in the book of Abraham which is contained in the Holy Book of Mormon.

The book of Mormon has its beginning in the Judeo-Christian Bible. It started with Joseph, son of patriarch Jacob. Joseph had two sons with his Egyptian wife. The sons were named Manasseh, the oldest, and Ephraim. When Joseph asked his father, Jacob, to bless his two sons, Jacob crossed his arms and placed his right hand on Ephraim's head and his left hand on Manasseh's head. Joseph protested. He knew Jacob's right hand should be on the elder, Manasseh. Jacob resisted.

Jacob told Joseph that Ephraim would be the founder of many nations, and Manasseh would be the founder of a great nation (America, USA). Jacob's prophecy came true with the rise of the British Empire's many nations "where the sun never set" and the founding of the United States of America. Pastor Herbert Armstrong, Sr., founder of the Universal Church of God, in Pasadena, CA, left the world a great legacy. During his life, he researched, preached, taught, and wrote the book, "Britain and America in Prophecy." He also wrote about the apostles chosen and commissioned by Lord Jesus Christ, where they traveled, found and taught the ten "Lost Tribes of Israel" and the Gospel.

Prophet Lehi and other Mormon prophets prophesied and inscribed on the Golden Plates that they married the daughters of Ishmael and sailed across the Great Waters (the ocean). Also, in the later days, a seer would be born with the name Joseph. All church members believe that Joseph Smith *is* that seer.

Lehi and others wrote on the Golden Plates, "We have the Words of the Hebrews and the Writings of the Egyptians" —Ancient Aramaic. In the Judeo-Christian Bible, there are Syrian relatives in the genealogy of Manasseh. Laban, which means white in Strong's Concordance, spoke Aramaic/Syrian and worshiped idols. The name Laban is found both in the book of Mormon and the Old Testament of the Judeo-Christian Bible. Present day Syrians, about 10%, speak modern Aramaic. The majority of present day Syrians speak Arabic. Arabic was derived from Aramaic.

Church of Jesus Christ of Latter Day Saints;
Joseph Smith Overcomes Bias, Clannishness,
and Opposition in the Church Toward
Semitic and Hamitic Peoples

From the founding of the Church of Jesus Christ of Latter Day Saints, the sacrament of universality was promised to all peoples of the world. The Golden Plates, with the Words of the Hebrews and the Writings of the Aramaic-speaking Egyptians, were loaned by God, initially, to two British Japhetic Americans, namely, Prophet Joseph Smith and his friend, Prophet Oliver Cowdery.

The Japhetic people of Britain and other European countries were accustomed to belonging to family clans, work clubs, and social cliques. People who didn't have anything in common with these members, such as race, religion, and work skills were not admitted to their groups. It was hard for "outsiders" to obtain admittance. The Japhethites and the Shemites of the twelve tribes of Israel were encouraged and commanded to marry and conduct business only in their tribes. Therefore, we see it was an innate trait to be clannish down through the centuries.

Joseph Smith encountered much clannishness and opposition from the Japhetic Americans during the early history of the Mormon Church in admitting Semitic and Hamitic (African) peoples. Joseph Smith acquiesced for a time until the church was larger and stronger. Eventually, the prophets and the church were blessed enough to throw off the darts and arrows of biases and misunderstandings. Today, the Mormon Church, known as the Church of Jesus Christ of Latter Day Saints (LDS), sends missionaries, two by two, to all countries in the world. People who are destined to study the teachings of the Judeo-Christian and Mormon Bibles are notified and contacted, sometimes, in mysterious ways. After three months of weekly lessons, students who grasp and understand the teachings and tenets of the Bibles and church are kindly invited to join the faith, be baptized, receive the Holy Spirit, study, and teach. This offer is extended to all races globally.

Polygamy, in a patriarchal society, is the marital practice of one man having two or more wives. Polyandry, in a matriarchal society, is the marital practice of one woman having two or more husbands. Lehi wrote on the Golden Plates, if we can paraphrase, "We have the words of the Hebrews. . ." and also Lehi and his family had the marital practices culture during the time the Golden Plates were inscribed. Polygamy was practiced. Jacob, father of Joseph, had twelve wives. David had several wives. Solomon, wise and rich, had many wives which he allowed to worship in their various faiths. When the Golden Plates were located in Palmyra, NY, around the 1840s, polygamy became an issue and practice for thirty years after the church was founded and found a permanent geographical home. Eventually, plural marriages were outlawed. However, in the Mormon book of Doctrines and Covenants, the marriage of one man to one woman was always decreed.

The building of the ark was a covenant between God and Noah. A covenant is a contract between two parties. One party agrees to do so much and the other party does the other half and completes the contractual transaction. Noah was commanded by God to build the ark which completed the physical part of the contract. God supplied the spiritual and miraculous aspects of the contract outside and inside the ark. It appears that the topography of the earth was a connected land mass before the deluge and up to the time of Peleg, which means divide, division, when the earth changed north and south poles polarity and divided into continents and islands.

God transported animals and creatures from this single land mass to the ark. Most likely, after the ark was completed, they entered the ark and were miraculously transformed. A faith teacher, whose congregational members have visited and taught the scriptures in almost every home in America and the world, suggested that God may have made the creatures that ended the ark very small, suspended animation, and/or reduced most of the animals and other creatures into a speck of DNA. These animals would not have to be fed during their voyage. Upon disembarking from the ark, God would restore the animals and other creatures to their normal sizes.

The ark was measured in terms of furlongs. One furlong was an eighth of a mile, 220 yards. On the outside, the ark measured five hundred feet long, seventy-five feet wide, and forty-five feet high with one window. However, inside the ark, its size could have been as large as a city block or larger. If so, God may have transformed space, time, and substance inside the ark. To get an inkling of this size expansion phenomenon, recall the British BBC drama *Dr. Who* and the TARDIS. The TARDIS is about sixteen feet around its four sides and about seven feet tall. Inside, it's enormous in size. God may have altered time inside the ark so that what occurred as days outside the ark may have been merely hours inside.

Genesis 8:4 says that the ark came to rest on Mount Arafat in the land today we call the nation of Turkey. After all of the floodwaters receded and Noah and his family and all the animals had disembarked, Noah planted a vineyard. The scriptures say that one day, Noah overindulged in drinking wine he had made from the grapes he had gathered from his vineyard; he got drunk and fell asleep. The average person prefers to drink with someone and not drink alone. Did Noah drink with his wife and she fell asleep too? We don't really know. However, when it comes to trying to understand the term, "Seeing your father's nakedness," the eighteenth chapter of Leviticus may shine some light on this subject. Whatever occurred, Noah cursed his grandson Canaan, whom he called his youngest son, and all of his descendants to the point of death for all of them. Also, Noah decreed that Ham would be a servant to his brothers Japheth and Shem. The connotative meaning of servant really means slave, bondsman. If this was the case with Ham's servitude to his brothers, then Ham could not have served as a priest to the people who was enslaving him. Consequently, all of Ham's descendants were denied the priesthood of service until the advent of Jesus Christ two thousand years ago. Jesus taught that He had come to free the captives and the captors; there was no slave, there was no master. All would be equal in the eyes of God. His kingdom would be a kingdom of priests in the order of Melchizedek as He is of the order of Melchizedek; the Aaronic dispensation having come to a closure at the crucifixion and resurrection of Jesus Christ. All Hamitic people, Cushites, Mizaimites, and Putites, the Libyans could now serve as priests. In addition, Jesus forgave all people of their infractions toward God and others. When Simeon of Cyrene helped Jesus carry His cross to Calvary, surely Jesus forgave him and all of his Hamitic kin. When Jesus, in agony, cried out from the cross, "Forgive them Father for they know not what they do." We all were forgiven of our past, present, and future infractions of not being able to hit the mark in our spiritual and secular life. Finally, as Jesus Christ was about to ascend to Heaven to be at the right hand of the Father, Jesus commissioned all of His disciples to go forth, spreading the Gospel of the Good News, and make disciples of all creatures in the world. This command by the Messiah included Japhethic, Semetic and Hamitic peoples worldwide.

Biblical scholars say that the world of all major religions believes that it took Noah 120 years to build the ark. This is not so. These biblical students refer to Genesis 6:3 where God says that "His Spirit shall not strive with man forever (720,000,000 million years) for he is indeed flesh, yet his days shall be 120 years (in present day time in years)." The ages of Adam to Shem approached a millennium, one thousand years which can be converted to 720,000,000 million years with a divine calculation as such: 1000 years x 360 x 2 x 1000 = 720,000,000 million years. Bible scholars and scientists are both right when they say one thousand years and 720,000,000 years respectively.

It has been reported on the Internet that at least one carpenter, using modern tools, built an ark with the same measurements as Noah's Ark in less than twelve years.

The true ages of the patriarchs from Adam to Shem may have been their ages multiplied by the divine calculation 360 and 2 but not by the 1000 in the math formula. Otherwise, they would be living with us today. How do we know they have not been "born again" and really living here with us? Most likely, they are now in Heaven.

We don't really know the complete age of Shem. The scriptures give us five hundred years but do not say, and all the years of Shem were such-and-such. Most biblical scholars, including the Persian prophet the glory of God, one of the greatest bible scholars of all scriptures of all major religions of the 19th century, whose interpretations and teachings are beneficial for mankind today and show great promise for the world in our near and far future, believe that Prophet Shem was Melchizedek, King of Salem.

The scriptures say that Melchizedek, "Without father, without mother, without genealogy, having neither beginning of days nor end of life, but made like the son of God, remains a priest continually." Re: Hebrews 7:3, Shem, like his relatives before him, was blessed with long longevity, approaching a millennium of one thousand years, a.k.a. known as 720,000,000 years, but never reaching this time period. However, when God decreed that man would only live 120 years after Shem, Shem may have continued to live many thousands of years while all his descendants and his brother's descendants may only lived 120 years maximum. Suppose Shem lived a total of eight hundred years and his prophetic years were calculated as follows: 800 x 360 x 2 = 576,000 years. Then if we divide 120 into 576,000 years, we get 4,800 generations of 120-year-old Noahic descendants during the 576,000 years life of Shem. Quite naturally, these 120-year-old descendants would believe, since Shem had lived so long, that Shem would have no father, no mother, no genealogy, and he was godlike living eternally. The only other divine person which could fit this profile would be Almighty God, no mother, no father, no genealogy. Jesus had a father: God.

A Brief Introduction to Plasmodium Malarial Disease

One of the greatest killers of present and past generations has been malaria. This disease kills between one to two million people yearly. It strikes the hardest in Sub-Sahara Africa, the Middle East, and Southern Asia. It is also found in places like South America and Central America. Presently, entomologists, biochemists, epidemiologists, and other related researchers have reported that there are at least 460 different mosquito species in existence and have been for more than 130 million years. One hundred species can act as vectors for malaria in almost any animal. The most common is the plasmodium gambiase which causes the malarial sickness in humans. We will discuss this species and the species which affect monkeys, birds, and other animals. Since animals of different species have a set of different chromosomes, we will try to show the adverse effect of the malaria parasitic ovum and its impregnation of red blood cells in spleen and liver. Ultimately, the eggs grow as protozoa in the circulatory cells of humans and other animals. Humans have forty-six diploid chromosomes and twenty-three haploid. The anopheles female mosquitoes have six diploid chromosomes and three haploid. First, we will list many malarial stages of the mosquito, oogenesis egg formation in the stomach of the mosquito during fertilization, and the protozoal stages in the red cells during asexual fertilization. Then we will attempt to show mathematically the gradual increase in chromosomal matter in the different malarial species' bodies during the sexual and asexual cycles in mosquitoes and animal hosts during various stages of eggs and protozoa development.

Human Diploid Chromosomes 2n	Mosquito Diploid Chromosomes 2n	Human + Mosquito Diploids 2n x 2	Mitosis Telophase 1	Meiosis Telophase 1	Meiosis Telophase 2 Macro & Micro Gametes

46	6.000	52.000	26.000	13.000	3.250 x 2
46	6.500	52.500	26.250	13.125	3.281 x 2
46	6.562	52.562	26.281	13.141	3.285 x 2
46	6.570	52.570	26.285	13.143	3.286 x 2
46	6.571	52.571	26.286	13.143	3.286 x 2
46	6.571	52.571	26.286	13.143	3.286 x 2
46	6.571	52.571	26.286	13.143	3.286 x 2

Note: After the fifth cycle, all calculations continue to repeat and stall. The reason for this phenomenon needs further study. However, during an esterification of equal moles of an acid reactant and an alcoholic reactant, the total reaction may stall. Additional alcohol moves the reaction.

Horse Diploid Chromosomes 2n	Mosquito Diploid Chromosomes 2n	Horse + Mosquito Diploids 2n x 2	Mitosis Telophase 1	Meiosis Telophase 1	Meiosis Telophase 2 Macro & Micro Gametes
46	6.000	52.000	26.000	13.000	3.250 x 2
46	6.500	52.500	26.250	13.125	3.281 x 2
46	6.562	52.562	26.281	13.141	3.285 x 2
46	6.570	52.570	26.285	13.143	3.286 x 2
46	6.571	52.571	26.286	13.143	3.286 x 2
46	6.571	52.571	26.286	13.143	3.286 x 2
46	6.571	52.571	26.286	13.143	3.286 x 2

Note: After the sixth cycle, all figures repeat and stall. During the titration of phosphoric acid, H_3PO_4, two hydrogen come off easily. The third hydrogen is very hard to remove from the PO_4, and the titration reaction stalls. It appears to be flat when charted.

Note: The ploidy chromosomal increase in the infected female anopheline mosquito diploidy may be due to aneuploidy mosaicism. Aneuploidy is a loss of one chromosome from a pair of chromosomes. Also, it can be the dislocation of a chromosome from a pair and travels to a pair of chromosomes, making it three chromosomes which are called trisomy. A gene having only one chromosome is called monosomy. In humans, deformities always result in individuals due to such a chromosomal dislocation. Partial aneuploidy occurs when only a fraction of another chromosome relocates with a pair.

If this is truly the case with the anopheles mosquito diploid increase after she bites the animal host, it may reveal what happens after the infected protozoa parasite is recycled through animal and mosquito hosts several times. The math in the present tables may show where the reduction of the malaria parasite cease, and the malaria medications are no longer effective. We know that the malaria meds are oxidized by the malarial parasites because the methyl anions (usually four) on the meds are lost and are attached to the malarial bodies. Some researchers claim that the meds are most effective when the meds lose the negative methyl anions, negative hydrides, and electrons have similar reducing effects.

Monkey Diploid Chromosomes 2n	Mosquito Diploid Chromosomes 2n	Monkey + Mosquito Diploids 2n x 2	Mitosis Telophase 1	Meiosis Telophase 1	Meiosis Telophase 2 Macro & Micro Gametes
48	6.000	54.000	27.000	13.500	3.375 x 2
48	6.750	54.750	27.375	13.688	3.422 x 2
48	6.844	54.844	27.422	13.711	3.428 x 2
48	6.856	54.856	27.428	13.714	3.429 x 2
48	6.857	54.857	27.429	13.714	3.429 x 2
48	6.857	54.857	27.429	13.714	3.429 x 2

Note: After the fifth cycle, all figures repeat and stall. These calculations are the same as the results of the human host diploid of forty-six chromosomes and the anopheles female mosquito of six chromosomes. The final haploid gametes of 3.429 is 0.429 higher than the A. mosquito microgamete of 3.00. This could mean that the zygote formed from 3.0 and 3.429 gametes would have a diploid of 6.429 and have abnormal properties. Gametes of 3.0 plus 3.0 would be "normal."

Throughout history, biologists and related scientists have discussed the malaria strain theory. This theory is based on the fact that viruses and bacteria mutate and form different strains as they infect and pass through various hosts over a period of time. These researchers believe that it is likely that anopheles and plasmodium malarias act in a similar manner. These mathematical models may show when changes take place in the anopheles and plasmodium mosquitoes. Note that pigs and cats have thirty-eight diploid chromosomes. Their gametes in the following model are between 2.0 and 3.0. This may mean that these gametes can't form zygotes because they are less than 3.0.

Pig Diploid Chromo- Somes 2n	Mosqui- to Diploid Chromo- somes 2n	Pig + Mosqui- to Dip- loids 2n x 2	Mitosis Telo- phase 1	Meiosis Telo- phase 1	Meiosis Telo- phase 2 Macro & Micro Gam.
38	6.000	44.000	22.000	11.000	2.750 x 2
38	5.500	43.500	21.750	10.875	2.719 x 2
38	5.438	43.438	21.719	10.859	2.715 x 2
38	5.300	43.300	21.715	10.857	2.714 x 2
38	5.429	43.429	21.714	10.857	2.714 x 2
38	5.429	43.429	21.714	10.857	2.714 x 2

Note: After the fifth cycle, the figures stall and level off; they are the same. Does cycle #1 mean that there is only one species when the female anopheles mosquito bites the malarial infected pig? Also, when the figures after cycle #5 are the same, is this where malaria drugs become ineffective? More study in this area is needed by malarial researchers.

Malarial scientists who have studied the fertilization of the male and female anopheles mosquitoes claim that they mate only once. The males fly into a swarm of females and deposit their sperms into a special pouch imbedded in the body of the female. From this pouch, throughout her fifteen day life cycle, she draws sperms to fertilize her eggs.

The male anopheles mosquito, similar to the male fruit bats, only eats nectar and sugary substances and rotten fruit containing a trace of fermented alcohol. The male anopheles mosquito is not a vector of malaria; doesn't bite host animals and doesn't consume blood. The female also eats nectar and sweet substances plus host blood. The female mosquito and the vampire bat may have a reduction/oxidation (redox) iron metabolism problem. If the blood iron in each case has lose electrons and oxidized to ferric iron +3, each human chromosomes combined.

Horse Diploid Chromo- Somes 2n	A. Mos- quito Diploid Chromo- somes 2n	Mos- quito + horse Diploid Chromo- somes 2n x 2	Mitosis Telo- phase 1	Meiosis Telo- phase 1	Meiosis Telo- phase 2 Macro & Micro Gametes

Ref: See page 35 for horse and A. mosquito diploid calculations

64	6.000	70.000	35.000	17.500	4.375 x 2
64	8.750	72.750	36.375	18.188	4.547 x 2
64	9.094	73.094	36.547	18.273	4.568 x 2
64	9.137	73.137	36.568	18.284	4.571 x 2
64	9.142	73.142	36.571	18.286	4.571 x 2
64	9.143	73.143	36.571	18.286	4.571 x 2
64	9.143	73.143	36.571	18.286	4.571 x 2

Human Diploid	Horse + Mosqui- To Dip- Loids	Horse + Mosqui- To Dip- loids + Human	Mitosis Telo- phase 1	Meiosis Telo- phase 1	Meiosis Telo- phase 2 Gametes
46	9.142	55.142	27.571	13.786	3.446 x 2
46	-- ?	6.893	3,446	1.723	0.431 x 2

Human would have to be reduced to ferrous iron +2 with electrons. What would happen, genetically, if a female anopheles mosquito, six diploid, would infect an animal host, such as a pig, thirty-eight diploid, and then bite and "infect" a human host, forty-six diploid? In the following table, we will attempt to derive an infectious scenario pattern using the previous mathematical modified model.

Pig Diploid Chromo- Somes 2n	Mosqui- to Diploid Chromo- somes 2n	Pig + Mosqui- to Chro- mosomes 2n x 2	Mitosis Telo- phase 1	Meiosis Telo- phase 1	Meiosis Telo- phase 2 Macro & Micro Gametes
38	6.000	44.000	22.000	11.000	2.750 x 2
38	5.500	43.500	21.750	10.875	2.719 x 2

Human Diploid Chromo- somes 2n	Mosqui- to + pig Chromo- somes 2n	Human + Pig + Mos- quito Chromo- ?	Mitosis Telo- phase 1 somes 2n x 2?	Meiosis Telo- phase 1	Meiosis Telo- phase 2
46;	44.000	90.000	45.000	22.500	5.625 x 2

Human Diploid Chromo-Somes 2n	Mosqui-to + Pig Diploid Chromo-somes 2n ?	Human + Pig + Mos-quito Diploid Chromo-somes 2n x 2 ?	Mitosis Telo-phase 1	Meiosis Telo-phase 1	Meiosis Telo-phase 2 Macro & Micro Gametes
46;	11.000	57.250	28.625	14.313	3.780 x 2
46;	7.156	53.156	26.578	13.289	3.322 x 2
46	6.645	52.645	26.322	13.161	3.290 x 2
46	6.581	52.581	26.290	13.145	3.286 x 2
46	6.573	52.573	26.286	13.143	3.286 x 2
46	6.572	52.572	26.286	13.143	3.286 x 2
46	6.572	52.572	26.286	13.143	3.286 x 2

Note: It appears that the combination of the above diploids would promote a malarial infection.

Horse Diploid Chromo-Somes 2n	A. Mos-quito Diploid Chromo-somes 2n	Horse + Mos-quito Diploids 2n x 2	Mitosis Telo-phase 1	Meiosis Telo-phase 1	Meiosis Telo-phase 2 Macro & Micro Gametes

Ref: See page 35 for horse and anopheles mosquito diploids.

64	6.000	70.000	35.000	17.500	4.375 x 2
64	8.750	72.750	36.375	18.188	4.547 x 2
64	9.094	73.094	36.547	18.274	4.568 x 2
64	9.137	73.137	36.568	18.284	4.571 x 2
64	9.142	73.142	36.571	18.286	4.571 x 2
64	9.143	73.143	36.571	18.286	4.571 x 2

64	9.143	73.143	36.571	18.286	4.571 x 2 *
Human Diploid	Horse + Mosquito Diploids	Horse + Mosquito + Human Diploids	Mitosis Telophase 1	Meiosis Telophase 1	Meiosis Telophase 2
2n	2n	2n x 2 ?			(con'd)
46	9.142	55.142	27.571	13.786	3.446 x 2
Human Diploid	Horse + Mosquito Diploids	Horse + Mosquito + Human Diploids	Mitosis Telophase 1	Meiosis Telophase 1	Meiosis Telophase 2
2n	2n	2n x 2 ?			
46	6.892	52.892	26.446	13.223	3.306 x 2
46	6.612	52.612	26.306	13.153	3.288 x 2
46	6.576	52.576	26.288	13.144	3.286 x 2
46	6.572	52.572	26.286	13.143	3.286 x 2
46	6.572	52.572	26.286	13.143	3.286 x 2

Note: The gametes derived from the calculation of the horse and A. mosquito appears to show partial aneuploidy because they are a fraction higher than haploids 3.0. Also, the diploids are higher than 6.0.

This is also true when the two diploids are combined with the human forty-six diploid. These results seem to promote a malaria infection.

Lord God Yahweh instructed Prophet Moses to teach the children of Israel that they were not to eat crustaceans and aquatic creatures and fish that had no scales. Commonly known crustaceans are shrimp, crabs, lobsters, crayfish, and scallops. Whales, catfish, octopus, and sharks have no scales and are frequently eaten by many different cultures worldwide. Most crustaceans, catfish, and sharks are scavengers and bottom feeders in the oceans and waterways. In the last two centuries, humankind has poured harsh chemicals, waste, and toxins into our oceans, lakes, rivers, ponds, and creeks. These bottom eaters have absorbed these poisons into their genetic systems. Case in point, take tuna which is known to contain elevated mercury levels. Our God knew this would occur before He created the heavens, universe, earth, moon, and the stars. Shame on us! Now it is our job to clean up our mess and God's environment. Listed below are some of the above aquatic animals, their chromosome diploids and haploids.

Aquatic Diploids Haploids Aquatic Diploids Haploids species

Species

Species	Diploids	Haploids	Species	Diploids	Haploids
Shrimp	254	132	Crayfish	>100	>50
Lobster	100	50	Carp	104	52
Crabs	208	104	Man	46	23

Aquatic Diploids Haploids Aquatic Diploids Haploids Species

Species

Species	Diploids	Haploids	Species	Diploids	Haploids
Salmon	54	27	Catfish	58	29
Shark	82	41			
Trout	58	29			

Note: Most people believe that sharks and catfish are similar in nature because they look alike and have scavenger feeding habits. However, catfish diploids are closer to most "common" fish than sharks. This may mean that farm-raised catfish are more compatible for humans than other wild aquatic scavengers.

It is not common for the above aquatic creatures to be infected by malarial mosquitoes because most of them don't have hemoglobin and red blood cells like most malarial hosts. Just to see what happens, we're going to subject shrimp to our mathematical mitosis and meiosis test model.

Shrimp Diploid	A. Mosquito Diploid	Mosquito + shrimp Diploids	Mitosis Telophase 1	Meiosis Telophase 1	Meiosis Telophase 2 Macro & Micro Gametes

Shrimp Diploid	Mosquito Diploid	Mosquito + shrimp Diploids	Mitosis Telophase 1	Meiosis Telophase 1	Meiosis Telophase 2- Macro & Micro Gametes
2n	2n	2n x 2			
254	6.000	260.000	130.000	65.000	16.250 x 2
254	32.500	286.500	143.250	71.625	17.906 x 2
254	35.813	289.813	144.906	72.453	18.113 x 2
254	36.227	290.227	145.113	72.557	18.139 x 2
254	36.278	290.278	145.139	72.570	18.142 x 2
254	36.285	290.285	145.142	72.571	18.143 x 2
254	36.286	290.286	145.143	72.571	18.143 x 2
254	36.286	290.286	145.143	72.571	18.143 x 2

Note: Shrimp have high cholesterol content. Shrimp can raise both the HDL and LDL cholesterol. A consumer who eats shrimp frequently adds to a genetic disorder known as hypercholesterolemia. This disorder is a mutation that is found on the LDL receptor on the nineteenth chromosome area.

Whales eat shrimp and often lose their direction and beach themselves. This could be due to a virus that some shrimp contract. God taught us the foods to eat: for Adam & Eve, fruits, nuts, and vegetables. After the deluge, God told Noah that he and his descendants could eat clean animals in addition to the Adamic diet. When Apostle Peter was worried about teaching non-Jewish gentiles, God sent a vision to Peter which contained a mixture of many animals descending from Heaven. God told Peter what He had made it clean. Primarily, this decree pertained to the teaching of the Gospel to all people of the earth.

Secondly, God would allow all people to eat foods according to their culture and availability.

Fertilization of the Female Malarial Mosquito

The male and female malarial mosquitoes pass through four stages from eggs to adults. The males emerge first from their watery birth sites. They form swarms and when the females are hatched, the females fly into the male swarms. The females are fertilized only once with male sperms. The females store the sperms in a part of their body called a spermathecae. The male returns to the swarm and fertilizes other virgin female malarial mosquitoes. Throughout the life cycle of the malarial female mosquito, she draws from the stored sperms in her spermathecae. In order to nourish her eggs and malaria protozoa offspring, it is necessary for her to draw blood meals from animal hosts.

When biologists conducted a genomic mapping of the plasmodium falciparum in the early 2000s, they discovered that this malaria had a fourteen chromosome diploidy as compared to other malarial species with a six chromosome diploidy and a haploidy of three. A fourteen diploidy is 2.33 times higher than a six diploidy. Could this increase in diploidy be one of the reasons P. falciparum is more virulent and deadlier than the six chromosome diploid species? In addition to the many tests and analyses that countless entomologists have applied to this diploid subject, let us subject the combined diploidy of the human diploid of forty-six and the fourteen diploid of the P. falciparum to our previous mathematical model analysis.

Human Diploid 2n	Plasmodim Falciparum Diploid	Human + P. Fal-ciparum	Mitosis Telo-phase	Meiosis Telo-phase	Meiosis Telo-phase
2n	Diploids	1	1	2	
2n	x	2	Macro	&	
Micro					
Gametes					
46	14.000	60.000	30.000	15.000	3.750 x 2

Human Diploid 2n	Plasmodium Falciparum Diploid 2n	Human + P. Falciparum Diploids 2n x 2	Mitosis Telophase 1	Meiosis Telophase 1	Meiosis Telophase 2 Macro & Micro Gametes
46	7.500	53.500	26.750	13.375	3.344 x 2
46	6.688	52.688	26.344	13.172	3.293 x 2
46	6.586	52.586	26.293	13.147	3.287 x 2
46	6.574	52.574	26.287	13.144	3.286 x 2
46	6.572	52.572	26.286	13.143	3.286 x 2
46	6.572	52.572	26.286	13.143	3.286 x 2

Note: The calculations for this test show a decrease in diploidy in the column for macrogametes and microgametes. For example, gametes initially 3.750 and finally 3.286 haploids are less than half of the haploid 7.000 for the plasmodium falciparum diploid fourteen. Malarial host with diploids below forty-six tend to show gametes less than 3.000 haploids, such as pigs, thirty-eight, and cats, also with a thirty-eight chromosome diploid. Most vegetables have diploids between eighteen and thirty. Tea is thirty, coffee between forty-four and eighty, penicillium 4.0; E. coli (Eschericia Coli) 1.00. Vegetable diets before, during, and after malaria infections may be better than eating high diploid meat diets.

Human Diploid 2n	Plasmodium Diploid 2n	Human + P. Falciparum Diploids 2n x 2	Mitosis Telophase 1	Meiosis Telophase 1	Meiosis Telophase 2 Macro & Micro

					Gametes
46	6.574	52.574	26.287	13.144	3.286 x 2
46	6.572	52.572	26.286	13.143	3.286 x 2
46	6.572	52.572	26.286	13.143	3.286 x 2

Note: The calculated macro and microgametes are less than the P. falciparum haploid, 7.00, diploid 14.00. Most host with diploids less than forty-six diploid human tend to show calculated haploids less than 3.00, such as pigs and cats of all kinds. Most vegetables have diploids between eighteen and thirty. Coffee ranges from forty-four to eighty, tea 30.0, penicillium 4.00, Eschericia coli (E. coli) 1.00. Pigs and cats have 38.00 diploids. A diet high in low vegetable diploids may be a better diet than a high diploid meat diet before, during, and well after all malaria infections.

Note: (3) When there is an increase in mosquito diploidy, it appears that a small percentage of fragmented chromosomes are dislocated from the animal host.

(4) The opposite appears to occur when there is a decrease in mosquito diploidy. A transfer of fragmented chromosomes may travel from the mosquito to the chromosomes of the animal host. However, small either transfer, a slight mutation might take place in the chromosomal genes and DNA of the mosquito and animal host. This mutation would be present in all of the complex formations in the mosquito sexual cycles and the animal host asexual cycles.

(5)The treatment of the malarial disease with known effective medications is a reduction reaction process. The mosquito eggs and protozoa are reduced when they receive negative methyl anions, negative hydride hydrogens and, of course, negative electrons from malarial medications and nontoxic compounds and natural substances containing this reducing substituent. When the malarial medications and substances that contain methyl anions, hydrides, and excess electrons, etc. lose this substituent (attachments), these core chemicals are oxidized. When other chemicals and malarial DNA components receive these leaving groups, they are reduced. Lose, it's oxidation; receive, it's reduction.

(6) There are many substances that have methyl anions and hydrides. Caffeine has four methyl anions on its fused six sided ring and its five sided ring. Caffeine may be an effective natural chemical, combined

with other essential malaria fighting drugs, that can help kill and cure malarial species. In addition, eosin methylene blue dye consists of three heterocycles rings with four methyl anions attached. Heterocycles are one, two, or three six sided rings with nitrogen, sulfur, or other atoms in their rings instead of all carbon atoms. Even the hemoglobin heme group with an oxidative/reductive iron atom in its center has four methyl anions in its periphery. It appears to be self-destructive for the female malarial mosquito to bite and suction blood from animal hosts when these methyl anions are harmful to her existence. That raises a question: is the female hungry for iron and can't obtain and metabolize iron in the malarial system?

(7) Most living organisms contain cytochrome P-450. This cytochrome neutralizes all toxins that enters the cells. It even neutralizes most medications, including malarial medications and cancer meds such as Taxol derivatives. This cytochrome may be the culprit which renders most malarial drugs inactive. Cytochrome P-450 might be effective in reducing the radioactive components in animals when they are exposed to various degrees of radiation. Radiation occurs because the radioactive element contains a disproportional amount of neutrons (gamma rays) to protons (alpha rays) and electrons (beta rays). The oxidative/reductive effect of the cytochrome P-450 might neutralize the protons and electrons, leaving the neutrons by themselves. This may "speed up" the radioactive half-life. Heterocyclic compounds such as eosin methylene blue (3, 7- Bis [dimethyl amine] pheazothionium chloride), bridge head carbons, metallic macrocylic bridgeheads, and cytochrome P-450 might act on radioactive components. Medical researchers report that bridgehead compounds slow the mitotic process in cells.

Malarial Ovum/Protozoa Cycle and Yeast Cycle Similarity

The malarial infectious cycle and the yeast growth cycle are similar in several ways. For instance, if we start at the reproduction stage where the malarial female macrogamete and the male microgamete join to form a zygote and the yeast cell joins two haploids of a female and male to form a diploid zygote, we find a similarity between malarial cycle and yeast. In the malarial oogenesis, egg forming stage, the zygote becomes an ookinete, and then an oocyst which grows very large until it is filled with countless sporozoites. These sporozoites are released into the female malarial salivary gland and then oocyst bursts. The female mosquito has two choices. She can either deposit these eggs into a still watery site where these eggs can rest on an egg raft and go through four stages before they hatch and become adults: male and female mosquitoes. Or the malarial female mosquito can bite a vertebrate host and deposit her sporozoite eggs into the host's bloodstream. Quickly, these eggs reach the host's liver and schizogony takes place. Trophozoites form in the red cells and the red cells form merozoites and each merozoite splits into a female macrogametocyte and a male microgametocyte is this asexual cycle. Then when the female malarial mosquito bites, the infected host and the macrogamete and microgamete join to form a zygote, the process starts over again. In the yeast growth cycle, the yeast often infects the host and/or forms a mature acus with several ascospores which are released for the germination phase. Next, the asexual budding of the two haploid cells occurs. In the female malarial cycle, this point would be where the female macrogamete haploid and the male microgamete haploid (one-half chromosomes) would eventually fuse together and form the diploid zygote. Then the meiotic oogenesis (egg forming) phase repeats.

Researchers who specialize in the study of yeasts, such as Candida albicans, have discovered that some yeast species can be treated with statin drugs. Most statin drugs chemical structures contain basic methyl anions the same as heterocylic malarial drugs. These drugs contain two or three fused benzoid rings, one or more nitrogen or sulfur atoms in place of the six carbon atoms that are normally found in benzene type compounds. It is possible that some statin drugs, in combination with other malarial drugs, might enhance the treatment of malaria.

Three ringed heterocylic compounds have been studied as a medicine to treat tuberculosis. It has been reported that INH (Isonicotinic Acid Hydrazide) and streptomycin have caused adverse effects in TB patients. Hydrazide is a salt. In its liquid phase as hydrazine, it is suspected to be carcinogenic. Streptomycin, on the other hand, affects the auditory nerve especially in young children.

Abortive Effects of Excessive Quinine and Excessive Caffeine in Vertebrates

There is a saying that excessive quinine will create a threat in abortion in pregnant females in their first trimester, second trimester, or their third trimester if their estrogen level drops below the normal levels. Quinine is a bi-benzenoid heterocyle compound that is used worldwide in combination with other drugs to fight malarial diseases. On its chemical structure core, there are two basic methyl anions and negative hydrides. The ring carbons have a positive charge. Caffeine has two negative methyl anions on the six benzenoid ring, and two negative methyl anions on the fused five ring structure along with several hydrides.

Northeastern European researchers have conducted studies and documented that there is a greater percentage of miscarriages in pregnant women who consume upward of 1000 mg of caffeine. The average cup of coffee and regular tea is about 200 mg. Green tea is a reduced tea and regular caffeinic tea is an oxidized tea. Green tea is picked and processed immediately and not allowed to become exposed to the air and moisture. Therefore, green tea retains its methyl anions and hydrides. There is a possibility that caffeine can be used to treat malarial patients and kill malarial eggs in mosquitoes and in the areas where they hatch.

Cytochrome P-450 System

Cytochrome comes from a Greek word meaning "cell color." A cytochrome is a hemoprotein, like the heme in hemoglobin. There are many cytochromes which include cytochromes A, B, C. The word cytochrome is shortened to CYP for better clarity, study, and differentiation. The CYP P-450 isoforms are unique because they can be activated and deactivated by substances that are either inducers or inhibitors. Lavinol in grapefruit juice is one example of an inducer (enhancer). This may be why doctors prescribe certain medications with the warning not to eat grapefruits or drink the juice while taking that medication. Cigarette smoke is another inducer. It has the tendency to induce the CYP 450 and increase the potency of the meds and other substances in vivo. Medical researchers claim that Taxanes and its derivative Taxol are strong inducers. However, researchers also claim that Taxol derivatives slow down the cell mitosis process in cells so that the ratio process of mitosis chromosome division and meiosis can get back in order. This probably applies for cancerous and noncancerous cells alike.

When doctors say that chemo kills healthy cells along with cancerous cells, what they mean is that all cells are affected and that is why some cancer patients lose weight until the chemotherapy is stopped.

Almost every substance is either an inducer or inhibitor that works on the CYP system. Dozens are listed in copyright literature on the Internet. Two CYP inhibitors that should be noted are anerrhoa carambola a.k.a. star fruit, and watercress. Star fruit is very sweet and delicious and elongated. It grows in most tropical areas globally. When it is sliced, the slices look like "perfect" five pointed stars. The star fruit contains many small amounts of vitamins and minerals, 12 mg of phosphorous, potassium, and 133 mg of magnesium (for reduction).

Watercress contains Ca, Na, Iron, K, vitamin C, phosphorous, plus other components and traces of Mg, Cl, and Sulfur (S). What is so significant about watercress is it contains between 1000 and 4,900 international units of vitamin A. Watercress grows in most areas globally. It is eaten as a health food. These two foods, when grown and eaten, could suppress the CYPP-450 in invertebrate and malaria egg sites. This procedure may prevent malarial drugs and other meds from becoming resistant. Also, malarial pesticides might be more effective and become less resistant. Everyone, including farmers and gardeners, should do more research and be encouraged to grow star fruit and watercress.

Water covered the whole earth. Moses wrote that the waters rose above the mountains and every living creature that was not in the ark was downed.

New Biblical Testament Commentary
and Many Mysteries Explained

How many Alphaeus were listed in the Christian New Testament? There was an Alphaeus the father of Matthew the Levi, one of the apostles of Jesus the Christ (Mark 2:14 KJV). Another Alphaeus was listed as the father of James. Another Alphaeus was listed as being married to a Mary, mother of James, Joseph, and Jude. Then on Sunday, the first day of the week, on the afternoon of the resurrection day, Alphaeus, a.k.a. Clopas, was walking with a Simon on the road to Emmaus which was seven miles from Jerusalem. The name Alphaeus means chief and leader. In biblical times, people had several names, either because their names could be translated differently into the many languages existing at that time or for protection from authorities.

It is noted in the Bible that grandfathers and fathers-in-law could claim their grandchildren and sons-in-law as sons. For example, in Eygpt, when Reuben saw his father's nakedness by sleeping with one of Jacob's wives, Jacob took Reuben's name from Rueben's tribe and bestowed it upon his son Joseph's two sons, Manasseh the elder and Ephraim the youngest. Jacob called this new tribe one-half Manasseh and one-half Ephraim. In a sense, Jacob claimed his grandsons as sons. Their mother was an Egyptian, the wife of Joseph.

In the biblical old testament in the book of Genesis, Noah referred to Canaan as "my younger son" when his intuition revealed to Noah that his son Ham had "seen" his father's nakedness. The curse would fall on Noah's grandson's descendants and not on Ham. Later, God would order Moses to kill all of Canaan's descendants for their evilness and vileness. Ham's other three sons were exempted from this curse of servitude and death. Ham's other three sons were founders: Cush-Ethiopia, Mizraim-Egypt, and Put-Libya.

Jesus's stepfather, Joseph, is the son of Jacob, a.k.a. James the Great. However, in Mary's genealogy of Jesus, her father is listed as Hali, a.k.a. Haliacum, and also listed as the father of Joseph, her husband. This is another case where the father-in-law claims his son-in-law as his son.

Alphaeus, father of Apostle Matthew,
father of Apostle James, husband of Mary,
who was the mother of James, Joseph, and
Jude; Alphaeus, a.k.a. Clopas, walking companion
of Simon on the road to Emmaus.

In every society, there are many people with the same name. This could have been the case with the name Alphaeus in the society of Jesus, a.k.a. Yeshua, his family, disciples, and friends. However, if the name Alphaeus listed in the New Testament belonged to only one man, we can make the following assumptions: First, it is odd that Alphaeus would be married to a woman named Mary with the sons James, Joseph, Jude, and possibly Simon. Could Alphaeus be Joseph, the stepfather of Jesus? There is no place in the Christian Scriptures that says Joseph had died and was not alive during the crucifixion.

It is possible that Matthew was the son-in-law of Joseph and was married to one of Jesus's stepsisters. Thus, Joseph claimed Matthew as his son as was the custom at that time.

Note: After the resurrected Lord Jesus walked with Clopas, a.k.a. Alphaeus, and Simon on the road to Emmaus, as explained in the scriptures, He broke bread and disappeared, the two rushed back to Jerusalem, a seven mile hike. It was important for Clopas to tell the eleven apostles, who were hiding in fear of the Jews, and the other disciples, who were observing the Jewish Siva. If Simon was the son of Alphaeus and Alphaeus was the stepfather of Jesus, then Simon may have been "Simon the Pharisee" whom Jesus said, "Simon, I have need of your home tonight." It may have been that brothers Simon and Joseph/Jose remained Pharisees until after the resurrection of Jesus. This may have been why the scriptures say that Jesus's brothers, possibly sisters, didn't believe in Jesus and his ministry and miracles. Brothers James and Jude may have been the Apostles that the scriptures refer to as James the lesser and Jude the son of James. Confusing?

There are two other names for Jacob: Israel and *James*. Jacob who could claim his grandsons as his "sons" may have been called by Jesus as James the greater and his grandson James the lesser. Using this claiming custom, Jesus could call Jude the "son" of *James*. Therefore, James the lesser and Jude the "son" of James, the grandfather, most likely were two of Jesus's apostles. Even so, these brothers may not have believed completely in Jesus until after the resurrection. Recall, these two brothers exclaimed, "Is he out of His mind?" when Jesus wanted them all to go to a certain area. James and Jude were afraid and expressed great anxiety. If James and Jude were not two of the twelve apostles, they could have said, "What does His traveling to 'dangerous' sites have to do with us? Let Him do His own 'thing.' Let Him and His apostles go where they want to go." If James and Jude were not involved, they most likely would not have expressed fear.

The Jewish Siva

The Jewish Siva consists of a total of forty consecutive days. It's divided into a set of three days, seven days, and thirty days. The Siva is observed in the home of the last home where the decedent abided. During the first three days, the relatives and friends of Jesus, the decedent sat quietly and mourned the death of their beloved who had died. Visitors only spoke if they were spoken to by the close relatives. No one went out or engaged in labor. Food and drink were brought in. In the case of Jesus, it was during the Sabbath weekend. Jesus observed his own three-day Siva; He died at 3:00 PM on Friday and did not appear certainly in their mist until after sundown at the beginning of Monday.

Thomas was not present when Jesus appeared to His disciples secluded in the locked room. Again, Jesus observed the seven-day part of His Siva by not appearing to the disciples until eight days had passed. At least one time, Jesus was hungry and asked for some food. The disciples, we believe, offered Him broiled fish and honeycomb. Thomas was present during this second visit. Jesus commanded the eleven apostles and the other disciples present to go to Bethany in thirty days and meet Him there. This would complete the forty-day Siva; three days + seven days+ thirty days = forty days.

The Gender of Apostle John?

John means Jehovah/Yahweh is gracious. Johanna also means Jehovah/Yahweh is gracious. Any word in the Bible that consists of Anna, Anne, or Hannah means grace. The personality of Apostle John was one of love. Until the book of Revelation, Saint John referred to himself as the apostle whom Jesus loved. He was the only disciple to write about the female disciple who anointed Jesus's head with expensive oil. Judas vehemently opposed this act of love. Jesus approved that woman's anointing, and in fact, Jesus said she should always be remembered and commended. John wrote about this incident in the Gospel of John.

John and Brother James, children of Zebedee and Salome, sister of Mary, mother of Jesus, were with Jesus along with Peter when Jesus conducted certain ministries when the other apostles were not present. Cousin John said that if all of the miracles and works that Jesus performed were written down, the world could not hold all the volumes. John and his brother James might have spent more time with Jesus as they were growing up and saw many miracles that occurred in the presence of Jesus. For example, grass where Jesus walked and flowers He may have touched grew much faster and more beautiful than others. He may have prayed and sick animals and people were suddenly healed. These acts may have just happened because of the divine nature of Jesus the son of the living God.

Cousin John and Cousin James loved Jesus so much that they asked their mother Salome to ask Jesus when He came into His kingdom that they could be at his right side. To ask Jesus such a question means that they were very close to Him. There is a saying, familiarity breeds contempt. While Jesus was on earth, He kept them close to Him.

In the Second Epistle of John, John wrote to the Church of Ephesus the following: "The elder, to the elect lady and her children, whom I live in truth, and not only I, but also all those who have known the truth." Verse 1, KJV

Verse 12, "Having many things to write to you, I did not wish to do so with paper and ink, but I hope to come to you and speak face-to-face, that our joy may be full."

Verse 13, "The children of your *elected sister* greet you." Dr. Isaac Asimnov, a brilliant scientist and writer, commented about the Second Epistle of John. He said the *elect lady* and *your elected sister* statements could be allegorical or literal, he wasn't sure. Dr. Asimnov was a scholar in Hebrew, Greek, and Egyptian theological and secular history. In fact, Dr. Asimnov is known for his *Commentary of the Old Testament* and *Commentary of the New Testament*; two books which are still in print and can be ordered through any bookstore that sells religious literature. Remarkably, he translates the names, persons, places, secular and religious customs, etc. where they appear in Hebrew, Greek, Egyptian, Latin, English, and various other Middle Eastern countries' writings

The first time a new Bible reader reads that Apostle John refers to himself in the Gospel of John in the third person as "the apostle that Jesus loved," one might wonder if Jesus loved Apostle John more than He loved the other apostles. Why did John write that statement? Was John trying to be different? Was he trying to hide something about himself? When Apostle John leaned on the breast of Jesus at the Passover a.k.a. the Last Supper, a custom in Jesus's time, the scriptures never show that the other apostles applied this familiar gesture when greeting or talking to Jesus.

Biblical historians of all faiths, Christian and non-Christian, point out that artist never paint Apostle John with a beard. They believe John was a teenager during the ministry of Jesus and at the Last Supper dinner. If John was a teenager, he would have been nineteen years or less at the commemorational Passover. When Jesus was twenty, Apostle John would have been about eight or nine. Could Jesus and John have been able to relate with such an age difference? Most unlikely. Therefore, John must have been in his middle or late twenties. However, John could have been older than Jesus. John the Baptist was six months older than Jesus. John the Baptist and Jesus were second cousins. Their mothers were first cousins, not sisters. Apostle John's mother and Jesus's mother were sisters.

Apostle John, "his" brother James, and Apostle Peter were always with Jesus. Jesus asked John to help prepare the Passover. Evidently, John had helped prepare the previous Passovers when Jesus was thirty-one and a half and thirty. John probably had a keen knack for preparing food affairs for the apostolic group. Now, this was the third Passover together, and Jesus was thirty-three and a half years old. He probably began His ministry in September when He was thirty. John the Baptist started his ministry six months earlier in the wilderness after Nisan 14, the Passover.

At the last Passover, Apostle John was probably seated at Jesus's right side, Peter to the right of John, and James to the left of Jesus. If Peter had been seated to the left of James, Peter would have had to reach over James and Jesus to ask John who was the traitor. The table must have been low, and they were seated on pillows. For John to have leaned on Jesus's breast, John would have had to straighten his right arm against the floor or another pillow. The entire seating for all present was probably more like a circle, not like Leonardo Da Vinci's painting of the Last Supper where Jesus and the ten apostles were seated on one side of a long table and one apostle at each end of the table. The seating may have been more like Salvador Dali's painting of the Last Supper.

Where was Judas seated? He was most likely seated across from Jesus. This way Judas could dip his bread into the soupy bowl between them as Jesus said the traitor would do so. The other apostles thought Jesus was sending Judas to buy more supplies since he was the treasurer when Jesus told him, "Go and do what you have to do and do it quickly."

When Jesus was arrested in the Garden of Gethsemane, John remained near Jesus during the unlawful trial and to the crucifixion on Calvary a.k.a. Golgatha, the place of the skull. The "a" at the end of Golgatha is the article "the." The word is Aramaic, and the article always comes at the end of a noun, like in Swedish, when one says the book, they say "buka."

In order to clarify Apostle John's gender status, let us review some references that pertain to elect, lady, and sister in Christian literature. Elect refers to a lady or the church, the Lord, Israel as a country, benevolent angels, members of the Christian faith, and Christian ministers: apostles, prophets, bishops, deacons, etc.

Lady in the Christian Bible refers to a female of a high position, someone considered as royalty. Also, it can be the church.

Sister refers to a Christian woman who is a member of the church. Also, a woman in the church who is similar in age to another member is called sister. A younger female is thought of as being a daughter in Christ. Older women are often called mother. The mother of the pastor of a Christian congregation is called the first mother and his wife as the first lady. A Christian congregation might refer to another congregation as a sister church.

On the road to Golgatha, Apostle John recalls the female disciples who followed Christ up that painful trail. He did not say the women and I followed Jesus. Did he include himself among the women. Did he blend in with them so well that he was indistinguishable? If so, was John's hair length below his shoulders? During the day, when he was working, did he wear it up in a bun in order to comply with the Jewish dress code that men should not wear their hair below their shoulders? Let's return to the courtyard when Jesus was being judged unlawfully by High Priest Joseph Caiaphas and his fellow cohort priests. John and Peter were present.

John wrote in the Gospel of John that the disciple whom Jesus loved knew the high priest. This high priest most likely would be Theophilus, one of the five sons of Ananus, who is mentioned in the book of Luke. Theophilus became high priest in AD 37 until AD 41. John seemed to not have any problem being present in the courtyard, whereas Peter was questioned three times by a maiden attendant as whether he knew Jesus. Peter denied the question three times. Peter spoke Galilean Aramaic with an accent. John also spoke Galilean Aramaic.

In 1986, Dan Barag and David Flusser published a document about the find of the ossuary/bone box of Jehohannah, a.k.a. Joanna or Yoanna, the granddaughter of High Priest Theophilus. Was this granddaughter the disciple whom Jesus loved? Was Theophilus the father of Zebedee, making him the grandfather of Apostle "John" and Apostle James?

Theophilus had two sons, Matthias, high priest in AD 65, and Yehohanan who was the father of Yehohanah, daughter and granddaughter of "Most Excellent Theophilus" whose non-honorary name may have been Zabeyah.

A Joanna, which means that God is gracious, stood near the Cross with the other female disciples. When Jesus saw His mother Mary He said, "Woman, behold your son." Then He looked at the beloved Apostle John and said, "Son, behold your mother." The scriptures say from that moment on that John took Mother Mary into his home. Biblical historians believe that John was very rich. We must not question the works and commands of Jesus. However, we can't help wondering why Jesus asked John to take care of His mother. Why John? God forgive us. Jesus had at least two sisters who may have married and left Joseph Thaddaeus and Mary Thaddaeus's household. There were four sons, James the lesser and Jude the son of James (his grandfather Jacob, Israel, and James). It appears that these two sons were apostles. Two other sons, Joseph, a.k.a. Joses, and Simon may have remained Pharisees until after the crucifixion of Christ. Then there was Alphaeus a.k.a. Clopas who may have been Joseph, the husband of Mary.

If Jude, son of Jacob (the greater), was the stepbrother and apostle of Jesus the Christ, and his surnames were Labbaeus and Thaddaeus, then Joseph, Mary, and their entire family would have these surnames. Because there are no "Js" in the Hebrew and Greek alphabets, all the names in the Bible starting with "Js" would be "Ys." For example, in Hebrew, Jesus would be Yeshua. In Greek, Jesus would be Iesous.

At the Cross, did Jesus declare the gender of Apostle John by calling him son? Was John a female whom Jesus accepted as a male in his ministry. Was this made plain to the other apostles? Is this why Jesus and the other apostles never called him John until Jesus addressed him in the book of Revelation? John identified himself as John, a male gender, in the book of Revelation. If this is true, perhaps Jesus did not want the world to know He had made this bold gender phenomenon at that time of the embryonic stage of the way of faith, later to be called Christianity at Antioch. Why is the gender of John so important? If Jesus accepted a female in His ministry as an apostle, then this act of love for women should settle any disputes today about whether ordained and sanctified women should serve as priests, ministers, and pastors.

Joanna, an Ancestor of Jesus the Christ and an Apostle, Kinsperson of Paul?

The name Joanna appears in two other places in the Christian Bible, also known as the Book. During the time of Jesus, mothers taught their sons up to the age of seven. Then the fathers took over the son's teaching. The genealogy that Joseph gave Jesus in the book of Matthew contained at least three names that had the same meaning as Matthew: gift of Jehovah. However, in the genealogy that Mary gave Luke, the name Joanna (Joannas) is listed. Joanna is a female name. Usually, only the male name in the genealogy is listed. Joanna is listed as the son of Judah, the son of Joannas, the son of Rhesa. . .etc. Was Joannas a woman and was given a male gender status?

Apostle Paul speaks of a Joanna who was an apostle, his kinsfolk who was in the faith before him. This reference was read in the King James Version but could not be found in the Strong's Concordance using key words. All of Paul's Epistles would have to be read in order to find this source. How many Joannas were there during the ministry of Jesus? Was Apostle John born with the name Joanna and was given the name John and his spiritual male gender at the Cross? If we truly knew the answers to these questions, it would help the church in allowing Christian women to be priests, ministers, and Pastors of Christian congregations. Amen.

The Crucifixion Day of Jesus Christ/The Great Day of the Lord

In Psalm 22 in the "Old Testament" of the Jewish and Christian scriptures, Prophet David prophesied about Jesus's crucifixion and resurrection eight hundred years before the Roman military started hanging people on the double bar/beam cross. Everyone, if they haven't, should read the 31 verse Psalm. The Psalm begins with the future Savior crying out, "My God, My God, why have you forsaken Me?" In Aramaic, "Eli, Eli, lama sabatah ni?" In Hebrew, "Eloi, Eloi, lama shabata ni?" "Why are you so far from helping Me?" On the day of Jesus's death, He said these very words. However, if Jesus felt that God the Father had abandoned Him, why did He, less than three hours later, say during the three hour of darkness prelude at the ninth hour, "Father, into Your hands I command my Spirit."

Is it possible when Jesus looked down from the Cross, He saw only one of His apostles—John, standing a distant with a small group of women disciples. Ten apostles, plus Judas Iscariot, were not present. They had abandoned Him, forsaken Him for fear of the Jews and the Roman soldiers. Our Father did not forsake Jesus, His disciples did.

Two OT prophets foreshadowed this great day of the Lord. Zephaniah (1:14–15) prophesied, "The great day of the Lord is near. It is near and hastens quickly. The noise of the day of the Lord is bitter. There the mighty men shall cry out."

> That day is a day of wrath, a day of trouble and distress, a day of devastation and desolation, a day of darkness and gloominess, a day of clouds and thick darkness. (1:15;71)

The prophet prophesied,

> Behold, the day of the Lord is coming

> It shall come to pass in that day that there will be no light. The lights will diminish.

> It shall be one day which is known to the Lord—neither day nor night but at evening time it shall happen, that it will be light. (Zechariah 14:1, 6, 7)

Jesus's body was in the tomb of Arimathea (means a height) only thirty-six hours, not seventy-two hours consisting of three twenty-four-hour sundown to sundown days. A twenty-three-hour sundown day to sundown day could be one day. Also, a twenty-hour sundown day to sundown day would be a whole day. Jesus's body was in the tomb three twelve-hour sunrise days; thirty-six hours, three whole days. In addition, His Spirit was in Sheol, a.k.a Hell, for thirty-six hours. The Pharisee rabbinical teachings about Sheol in the time of Jesus's ministry consisted of three sections: Abraham's bosom, Turturos, and a great unbreachable chasm. Abraham's bosom was called Paradise. Turturos was where bad sinful souls were tortured with unbearable heat and pain. No souls could cross the chasm to the garden of Paradise. Remember the sinful rich man who died and descended into Hell. He saw Lazarus, the humble beggar, in Abraham's bosom and asked God to let Lazarus dip his finger in water and place a drop on his tongue. God said it was too late for the rich man. And God wouldn't even let Lazarus warn the rich man's brothers, so they would repent and avoid going to Hell.

So when the kind thief on the cross, next to Jesus, asked Jesus to remember him when He came into His kingdom, Jesus said to the thief, "I tell you, today you shall be with me in Paradise." A change in the syntax could read, "I tell you today, you shall be with me in Paradise." However, during the thirty-six-hour interlude, neither Jesus nor the repenting thief ascended to Heaven. They went to the Paradise in Hell—Abraham's Bosom. This was confirmed early Sunday morning after Jesus was resurrected. Jesus said to Mary Magdalene, "Touch me not, for I have not ascended to my Father." Jesus did not go to Heaven during His thirty-six-hour entombment and His glorious visit in Hell.

From the Cross, Jesus took on the sins and adverse forces of all humanity. He said, "Forgive them Father, for they know not what they do." If Jesus thought that the Father had forsaken Him, He would not have asked God to pardon everyone living on earth at that time, and hopefully, all people born to the present era.

Another declaration Jesus made while hanging and suffering on the cross was when He saw His lone apostle, John, and His mother, Mary. The scriptures read, "Now there stood by the cross of Jesus His mother (Mary) and His Mother's sister (Salome, mother of John), Mary the wife of Clopas, a.k.a. Alphaeus, a.k.a. Joseph the stepfather of Jesus, and Mary Magdalene. Clopas and his wife Mary had sons named James, Joseph, Jude, and Simon; re: Clopas and Simon on the road to Emmaus. (*See* John 19:25.)

Verse John 19:25 would be more explanatory if the syntax were changed to the following: "Now there stood by the Cross of Jesus, His mother, *Mary the wife of Clopas*, and His mother's sister (Salome), and Mary Magdalene.

In verse John 19:26 KJV, when Jesus therefore saw His mother (Mary) and the disciple whom He loved standing by, He said "Woman, behold your son (referring to Apostle John). Son, behold your mother (referring to Mary)!" The scriptures say from that hour on, John, the disciple whom Jesus loved, took Mary into his home. The home that Peter and John, the other apostles, and disciples were in while they were observing the three-day part of the Siva of Jesus must have been the home of Apostle John. We should not question the commands and actions of our Lord Jesus Christ. However, God forgive us, some disciples might wonder why Jesus made His mother a ward of Apostle John. Throughout the centuries from the crucifixion and resurrection of our Lord Jesus to the present era, church historians claim that artists have never painted Apostle John with a beard; they believe that he was a teenager.

This would have made John twelve years, or less, younger than Jesus. Could it also be that John was a female and at the cross when Jesus called him "son," Jesus gave John a clerical male status? If John were a teenager, that would have meant that John was about eighteen years old and Mary was in her midforties. John must have been older, near the age of Jesus because they grew up together.

Why would Jesus command Apostle John to care for His mother after His crucifixion and resurrection? The scriptures say that Jesus had four stepbrothers and at least two stepsisters. Nowhere in the scriptures does it say that Joseph, the stepfather of Jesus, was dead during the crucifixion event. Were the stepsisters of Jesus married and had departed from the Joseph Thaddaeus's household? The scriptures say that Matthew was the son of Alphaeus. How many Alphaeus were there? Does the name Alphaeus refer to only one person in the New Testament? If so, then Matthew may have been married to one of Jesus's stepsisters. If this was the case, then Joseph, a.k.a. Alphaeus, claimed Matthew as son because he was Joseph's son-in-law. Recall Jacob, grandfather of Ephraim and Manasseh, called "sons;" Noah, grandfather of Canaan, called "my youngest son;" and Joseph, son-in-law of Heli, father of Mary, called "son" in the genealogy given by Mary to Luke.

So what does all of this mean? It means that Jesus commanded Apostle John to care for His mother for a certain time while she was alive. When John wrote to the church of Ephesus, was Mary in John's care? When John was banished to the island of Patmos, Mary was not with John. John was nearly ninety years old. Mary would have been more than one hundred. Church historians say that at the time of the diaspora, a.k.a. the Jewish Dispersion, or before, Mary Magdalene took the mother of Jesus to Gaul, a.k.a. France. Are her burial remains in Lourdes, France, north of the Pyrenees Mountains, dividing France and Spain? Is her holy spirit capable of descending from Heaven to earth? Is this why countless people of faith frequently observe visions of Mother Mary at the Grotto Springs in Lourdes? In addition to the visions of Mother Mary, it is proclaimed that the spring water, the grass, the evergreen trees, the soil, and the forest air are conducive for healing bodies and souls. In His holy name we pray. Amen.

On the day of the crucifixion of Jesus, many strange phenomena occurred. Time, for example, seemed to change and be extended. From the start of the Passover, during the "trial" of Jesus, His torture and crucifixion, so many things happened that would not occur in a "normal" day. Sunset in Israel at Nisan 14 occurs around 5:00 PM. It appears that sunset didn't occur until around 9:00 PM. Recall the prophets who said on the day of the Lord in the evening, there would be light and during the day, there would be darkness. Jesus's body was entombed nearly thirty-six hours and his spirit and soul were in Sheol for thirty-six hours. These were three twelve-hour sunrise days, not sundown days. Jesus gave up His Spirit at the ninth hour—3:00 PM.

The time it took for Joseph of Arimathea to go to Pontus Pilate and ask for the dead body of Jesus was seemingly a long time. When Joseph of Arimathaea informed Pilate that Jesus was dead, Pilate was surprised. Pilate believed that Jesus was the King of the Jews. Otherwise, he would not have inscribed a placard to be placed over the head of Jesus after He was nailed to the cross which said He was against the protests of the incriminating Jews. Also, Pilate washed his hands of the crucifixion in public which indicated that the death of Jesus was on certain Jews for generations to come.

Perhaps Pilate believed that Jesus would free Himself from the cross and exonerate him, his wife, who may have also believed that Jesus was King of the Jews. Upon hearing that Jesus was dead, Pilate sent word to the centurions on Golgatha to confirm of His death. The custom to confirm or hasten the death of a prisoner on the cross was to brake both legs. This way, the prisoner wouldn't be able to breathe and would suffocate. However, when Centurion Longious came to Jesus, he could see that He was dead. To confirm this, the centurion pierced Jesus' side with his spear and outflowed water and blood. This information was quickly relayed to Pilate. Thus Pilate gave Joseph of Arimathaea permission to retrieve the bloody body of Jesus the Messiah. Pilate must have known Joseph of Arimathaea because a Roman governor would not have released a "state prisoner" to a stranger. A stranger might have done various things with the bloody body of Jesus whom Pilate had declared publicly on a wooden placard: JESUS CHRIST, KING OF THE JEWS in Hebrew, Greek, and Latin. This transaction of communication must have taken more than one-half hour. And sundown was around 5:00 PM. The age old dispute how Jesus was nailed to the cross might be clarified by how He was removed from the cross.

Some biblical scholars have argued that Jesus was nailed to a dogwood tree. Others have vehemently taught that Jesus was nailed to a single vertical stake with no crossbar or beam. Still, others believe that Jesus carried a cross with a vertical slender structure with a permanently attached crossbeam. These lateral believers believe that Jesus struggled with this type of structure from the prison up to Calvary with Simean of Cyrene at one point assisting Him.

Josephus, the great Jewish historian, wrote that wood was scarce in Jerusalem and once invading soldiers had to go ten miles to obtain enough wood for seize. Therefore, it is possible that a cross structure with an attached crossbar stood permanently in the ground, and prisoners were nailed or tied to another crossbar which could be hoisted up in less than ten seconds with ropes or chains and tied to the tilted short stake in the ground behind the cross. When we try to picture how the body of Jesus was removed from the cross, we realize how He was impaled and hoisted up on the cross.

His removal from the cross would be the reversal of His impalement. Joseph of Arimathea and Nicodemus most likely would have spread a long burial cloth on the ground. Then they would have requested a centurion to remove the nails from His feet. Slowly, the centurion(s) would have lowered the crossbar holding Jesus's hands. Then the centurion(s) would have removed the nails from Jesus's bloody hands as Joseph and Nicodemus embraced His body and laid Him on the burial cloth. This action may have taken more than one-half hour. Sunset was about 5:00 PM. Most likely, the female disciples didn't touch the body of Jesus. If they did, according to Jewish law, they would not be clean until evening. The scriptures say that a group of women followed the body of Jesus, probably on a death litter, to the new tomb of Joseph of Arimathea, but they didn't go in. This probably took more than one-half hour. Sundown was around 5:00 PM. Also, the scriptures say that Nicodemus brought between seventy-five and one hundred pounds of spices to anoint the body of Jesus. According to the Jewish burial custom, a deceased body is smeared with burial spices and wrapped from neck to feet with strips of white linen. It must have taken at least an hour to render this sacred sacrament to the body of Jesus. By this time, on a regular day, sunset would have occurred. Christ, being the sacrificial lamb, would have had His body anointed before the Passover day ended and the Sabbath day evening began.

The burial spices used were aloe, myrrh, and frankincense. Apparently, Joseph of Arimathea and Nicodemus did not go to the home where the eleven apostles, Mary, mother of Jesus, Mary Magdalene, and the other disciples were gathered observing the three-day Siva. The women went home to prepare spices to anoint the body of Jesus on Sunday, the first day of the week. They could not perform this deed on Saturday, the Sabbath. It would have been an act of laborious work which they could not do on the Sabbath. Therefore, the women did not know that the anointing and wrapping of the body of Jesus had been completed. The day with light may have been extended between 5:00 PM and 9:00 PM so that the anointing could be completed before the Passover day ended.

Again, did Pontius Pilate know Joseph of Arimathea well enough to entrust him with the battered body of Jesus the Christ, "King of the Jews?" Biblical historians claim that Joseph of Arimathea worked for Pontius Pilate. Joseph, who was very rich, had ships with which he transported tin ore from the mines of Wales in the British Isles to the iron workers of Pilate. Legend has it that Joseph took Jesus several times to "the isles of the sea" when Jesus was a youth. Secular historians have documented that there were tin mines and Israelites in Briton before Jesus was born. The Roman governors knew all of the Pharisees and Sadducees priests. In fact, in AD 67, several priests were sent to Rome and put on trial. Flavius Josephus, the Jewish historian AD 37—100, wrote that he sailed to Rome to defend these accused priests. Pilate knew the works of Jesus, His followers, and ministry. That is the main reason Pilate released the body of Jesus to Joseph of Arimathea.

Let us try to imagine the temperament and mind-set of Pilate from the time of Jesus's "trial" until His death on the cross at Calvary. Pontius Pilate must have been furious and extremely angry. The accusers of Jesus had trapped Pilate by proclaiming that Pilate was not a friend of Rome if he did not order the death of Jesus in place of zealous Barabbas. Pilate's wife admonished Pilate not to get himself involved in this mock trial of Jesus. Pilate tried to adhere to the warning of his wife and sought to release Jesus, but the chief priests maneuvered Pilate into a political corner with only one outlet—the crucifixion of Jesus by saying Pilate was not a friend of Caesar if he allowed Jesus to live. Against his will, Pilate ordered Jesus to be scourged. The Roman soldiers pressed a crown of thorns on the head of Jesus, mocked Him, spat upon Him, and hit Him with their palms and fists. Pilate referred to Jesus, "As this just person."

As the extremely fatigued Jesus struggled up the road to the place of the skull, Pilate may have paced the marble floor, sat at his marble desk, stared at the floor and sometimes into space. Pilate must have been waiting for some miracle to take place. Would Jesus, "King of the Jews" escape the wicked cross, he may have wondered. Judas may have thought if he could betray Jesus and place Him in a precarious position, Jesus would have to use His mighty powers and smite the Roman military and all of the enemies of the Jewish people, come to power, and rule the world. When Jesus failed to do that, Judas must have felt lost with an unbearable pain in his heart and being. He also wanted to die; he tried to kill himself. Pilate must have felt an unbearable pain when it was confirmed that Jesus was dead. With all the commotion concerning Jesus's trial, Pilate and others probably didn't ate breakfast neither lunch. When Jesus was driven to the cross, Pilate may have refused all food and drink brought to him by palace servants. He fasted as Jesus suffered. Upon hearing the death of Jesus and the release of His scarred body to Joseph of Arimathea and Nicodemus, Pilate may have whimpered and tried to cry; something he had not done since his youth. Pilate's wife, who was always near his side, may have embraced his head to her chest and then led him to their bedroom where he may have collapsed, tightened himself into a fetal position, and sobbed with hands covering his trembling face.

It appears that the timeline of the lateral Passover day may have been altered and extended. Out of chaos, God loves order, unity, and symmetry. God may have shortened three twenty-four-hour sundown days into three twelve-hour sunrise days; a total of thirty-six hours instead of seventy-two hours. From 3:00 PM Friday afternoon to 3:00 AM Sunday morning, it would be a total of thirty-six hours. Several timeline scenes of sights and sounds may be as follows:

Prelude

Date	VWC		Hours		
Friday	12:00 PM	to Friday	3:00 PM	Darkness	3

Interlude Scenario #1

Friday	3:00 PM	to Friday	9:00 PM (?)	Light	(6
Friday	9:00 PM	to Sat.	3:00 AM	Darkness	6
Sat.	3:00 AM	to Sat.	9:00 AM	Light	6
Sat.	9:00 AM	to Sat.	3:00 PM	Darkness	6
Sat.	3:00 PM	to Sat.	9:00 PM	Light	6
Sat.	9:00 PM	to Sunday	3:00 AM	Darkness	* 6) = 36

*Possible resurrection from Sheol back to earth plane

Postlude

| Sun. | 3:00 AM | to Sun. | 6:00 AM | Light | 3 |
| Sun. | 6:00 AM | to Sun. | 5:00 PM | Light | 11 |

Scenario #2

Light and darkness may have alternated every hour from Friday 9:00 PM to Sunday 3:00 AM.

Scenario #3

Light and darkness may have alternated every minute from Friday 9:00 PM to Sunday 3:00 AM.

Light and darkness and VWC continued on next page. Scenario #4

Scenario #4

Light and darkness may have alternated every other second from Friday 9:00 PM until Sunday 3:00 AM when the Messiah Jesus the Christ's entombed and linen-wrapped body received His resurrected spirit from Hell. A flickering alternating light and darkness effect would have manifested this phenomenon as a scary kaleidoscopic event.

Scriptures around the world whose prophets have had access to the crucifixion time with the power of the Holy Spirit have recorded that the whole earth shook with the effect of a giant earthquake. During this period, there may have been thunder, lightning, hail, cold rain, and kaleidoscopic light.

The Christian scriptures say that tombs were opened and dead people came to life and walked the scary streets. No wonder the temple guards were not at the tomb when Jesus was resurrected. If the Roman soldiers were guarding the tomb and ran away, they would have faced death for leaving their post. Three AM for the Roman soldiers was the cockcrow hour to change the guards. The winter hour was forty-five minutes in winter. In the summer, in June, the hour was seventy-five minutes. On Nisan 14, the hour was about sixty minutes.

Origin of the Holy Name Jesus

Church scholars claim that for centuries, the Romanized letter "I" was replaced with a "J" at the beginning and end of certain words created with the English type alphabet. Hebrew and Greek alphabets don't have a "J." Jesus's name in Hebrew is Yeshua and Yoshua. When the Hebrew scriptures were translated into Greek, the writers substituted "I" for "Y." Jews are not allowed to say Yahweh, they say Yehweh. In Greek, Yeshua was written as Iesous. Sous is feminine. When the Greek scriptures were translated into English type words, Iesous was written as J-e-s-u-s, one of the most beautiful, powerful, and healing names in the heavens and universe. Is it possible that God has allowed us to name and claim Jesus? In the book of Revelation, Jesus says to Apostle John, "And I will write on him my new name." This new name will be written on the thigh of the returning Messiah. (*See* Rev. 3:12 KJV.)

The Name(s) and Titles of God

Jehovah * Yahweh * I Am * Lord God * Mighty God * Heavenly Father * Lord of Hosts * Eternal God * Everlasting God * God of Heaven * Holy One of Israel * Almighty God *

The Name(s) and Titles of Christ

Origin of the Holy Name Allah

The Arabic language is derived from Aramaic. In middle Aramaic, west of the Euphrates River, the name for God was Aloha. The "a" at the end of the word is the article "the." In Swedish, the word for the book is buka. Aloha, that is the name in Tahitian and Hawaiian for hello, good-bye, and love. How many times do we say that God is love? And that God loves us?

East of the Euphrates River, the Aramaic word for God was Alaha. When Alaha was Arabized, a grammatical mark shaped like a "w" was placed on top of the "L." This mark is called a shadda. Anytime a shadda is placed on top of a constant, such as an "L," it doubles that constant. Therefore, Alaha would read Allaha. Remove the article "a" for "the" and we have A-L-L-A-H.

A voltaic battery has two electrodes, an anode and a cathode, a positive and a negative. They work together to pass electricity through the battery. We must all work together to glorify God and do His will. Amen

September, the Original Birthday Month of Jesus Christ

On Nisan 14, the Passover date, Jesus Christ was thirty-three and six months, thirty-three-and-a-half years. Prophets Daniel and Isaiah were inspired by God to write that the Messiah would be cut off in the middle of the week. A week consists of seven days. A half of a week would be three-and-a-half days which Ezekiel decreed would be three-and-a-half years.

Before and during the time of Jesus, Jewish priests started their ministry at the age of thirty. John the Baptist started preaching six months before Jesus. John was six months older than Jesus. The Jewish month date Nisan 14th falls between March fifteenth and March thirtieth. If Jesus were thirty-three and six months on Nisan 14th, in order to find the English month when Jesus was just thirty-three, we count backward six months. That last month falls in September. The Catholic and Greek Orthodox Church(s) historians claim that December ninth was the Day of Conception of Mother Mary. When we count forward nine months, the ninth month falls around September ninth, more or less. Therefore, Jesus's birthday is between September ninth and September thirtieth. The climate in Israel was warm for Mary and Joseph to travel to Bethlehem and for shepherds to be out on the hills with their sheep. Israel is very cold on December twenty-fifth. Most prefer celebrating Christmas in December.

Book Highlights and Commentary

1. In the formative years of the creation of the earth, it was predominantly water. Moses wrote about this aspect of creation in the book of Genesis 1:2 (NKJV), "The earth was without form and void and darkness was on the face of the deep. And the spirit of God was hovering over the face of the waters."

Peter also wrote about the earth being created out of water in the Second Epistle of Peter 3:5 (NKJV), "For this they willfully forget that by the word of God, the heavens were of old and the earth standing out of water and in the water."

God demonstrated to the children of Israel in the desert that He could turn rock into water and water into rock. God instructed Moses to speak to the rock, but Moses struck the rock with his staff and out came water." NKJV

These miracles raise the possibility that when God divided the Red Sea, He turned a pathway into a rock, so the children of Israel could crossover on dry land. The Bible says that the waters jellied. We often think of gelatin and jell-o jelling. This is a soft jell. The wide pathway across the Red Sea was a hard jell like stone. If this occurred, the structure would have been like a sheet of ice on a lake with some aquatic creatures swimming around beneath the pathway stone layer, above the sand on the sea bottom. When God turned the stone back into water, the Egyptians most likely became bogged down in the wet sand and sediment. Some Christian Bibles say that God killed all of the Egyptian livestock. Then the Egyptians would not have had horses and chariots. They would have been on foot. Some Bibles say that God only killed the Egyptian cattle. Therefore, they could have been in chariots and on foot. However, when the heaps of water receded and rushed to fill the roadway void, the Egyptians, being bogged down and could not escape, drowned.

More Water References:

In the time of Jesus Christ, there were three main religious sects: Pharisees, Sadducees, and Essenes. The Essenes' main site was the Qumran Community. They were noted for their pious lifestyle and immersing themselves twice a day in water. Biblical scholars are not sure whether Jesus Christ was an Essene. Jesus did allow himself to be anointed with oil. Essenes didn't anoint themselves with oil. When the female follower anointed Him, Judas Iscariot protested. He said that oil was very expensive and it could have been sold for much money. In addition to Jesus admonishing his disciples to observe the Passover each Nisan 14, He said to remember and commemorate that female's act of love. John/Joanna (?), the disciple that Lord Jesus loved, wrote about this pre-crucifixion event in the Gospel of John. He was the only apostle to do so. John was associated with love.

In the Second Epistle of John, he writes as follows: The elder, to the *elect lady and her children*, whom I love in truth and not only I but also all those who have known the *truth* because of the truth which abides in us and will be with us forever. (*See* 2 John 1:1–2.)

The children of your *elect sister* greet you. Amen. (*See* John 1:13.) Webster Dictionary and Strong's Concordance say that the terms *elect lady* and *elect sister* can refer to either the church or a female of royal status.

When Jesus cried out from the cross, "My God! My God! Why have *you* forsaken Me?" If the noun *you* was singular, Jesus may have been speaking of God. If the *you* were plural, Jesus may have been referring to His eleven apostles who were hiding because they were afraid of the Jews and, of course, the Roman Soldiers. In addition, the ten apostles, minus Judas Iscariot, close disciples, and family members could have gathered at the home where the deceased Jesus last abode to start the three-day Siva, followed by the seven-day Siva, and then the thirty-day Siva. A total of forty days.

Apostle John was the only apostle to follow Jesus to the crucifixion site, Golgotha, the place of the skull. John wrote in his Gospel that a small group of female disciples followed the battered Jesus up to Calvary. He didn't include himself as the disciple that Jesus loved.

Neither did he include himself when he wrote about the women who stood at a distant near the cross. However, Jesus called out to him without calling his name, "Son, behold your mother. Woman, behold your son." If John were a woman standing among other women, Jesus may have declared his status as a man. Why did Jesus single out John to take care of His mother? Jesus had four stepbrothers and at least two sisters. Alphaeus, a.k.a. Clopas, may have been Jesus's stepfather, Joseph, husband of Mother Mary. Nowhere in the Bible does it say that Joseph had died. John means Yahweh/Jehovah is gracious, so does Joanna.

1. Then Jesus came from Galilee to John (the Baptist) at the Jordan (River) to be baptized by him. (*See* Matthew 3:13 NKJV.)

"When He had been baptized, Jesus came up immediately from the water, and behold the heavens were opened to Him, and He saw the spirit of God descending like a dove and alighting upon Him" (Matthew 3:16 NKJV).

"And suddenly a voice came from heaven saying, this is My beloved son, in whom I *am* well please" (Matt. 3:17 NKJV).

John 2:1–10 NKJV: Mary told Jesus the wedding guests had run out of wine. Jesus rebuked her. However, Jesus told the servants to fill several five gallon jars with water, draw out a portion, and take it to the governor of the feast. The governor, when he had tasted it, exclaimed it was *new wine*.

Methods for Making Wine: (1) Water, juice, and baker's yeast

(2) 2Methane > 1 Ethene + HOH = Ethyl Alcohol + HOH (3) 2 CO_2 + 3HOH + Metallic Catalyst + pressure = CH_3CH_2OH + 6 OH > {6HOH+6OH} Oxygen and Hydrogen Bonding. "Heavenly" made Ethanol and wine; this is a theory.

2. Confirming the ages of the universe and earth, researched and copyrighted by astrophysicists and geophysicists; decoding and correcting the theological age of the earth by an advanced equation revealed by God, the Eloheem, using the constant pi (3.14) and derivatives of pi (3.0, 3.1) to convert the Gregorian year from the biblical year by dividing it by a form of pi, decoding the prophetic one thousand years, a.k.a. a day with the Lord in Heaven, a.k.a. *forever*, results to 720,000,000 million years equals one creative day that is 1000 x 360 x 2 x 1000 = 720,000,000 years.

Geophysicists claim that the earth is at least a coded six thousand years old. We are presently in God's rest day, the seventh creative day of God's creative calendar. It was reported by theologians that in 2014, the biblical year and existence of humankind on the earth was 5774. This figure of 5774 says that we are still in the sixth day. When we decode this figure, we get the following: 5774 x 360 x 2 x 1000 = 4,157,280,000 billion for age of the earth. This result x 3.0 = 12,471,840,000 billion years for the age of the universe. Scientists have reported that the universe is approximately 13,700,000,000 and the earth is one-third of that figure, 4,566,666,667 billion years. When we calculate and decode 6,000 years, we get the following results: 6,000 x 360 x 2 x 1000 = 4,320,000,000 billion years for the earth and 3.0 x 4,320,000,000 = 12,960,000,000 billion years for the universe.

It's obvious that these figures must be higher. When we codify this universe figure back to the biblical and Gregorian years, we get the following results: 12,960,000,000/divided by 360 = 36,000,000/2 = 18,000,000/1,000 = 18,000/3.0 = 6,000 biblical years divided by 3.0 = 2,000 Gregorian year. The universe age figure is much higher. When we codify 13,700,000,000 billion years, we get the following results: 13,700,000,000/360 = 38,055,555.56/2 = 19,027,777.78/ 1,000 = 19,027.778/3.0 = 6342.593 biblical years divided by 3.0 = 2114.198 Gregorian year. However, when we divide 6342.593 by pi, we get the following results: 6342.593/3.14 = 2019.934. When we increase the universe figure to 13,750,000,000, we get the following results: 13,750,000,000/360 = 38,194,444.44/2 = 19,097,222.220/1000 = 19,097.222/3.0 = 6,365,741/3.14 = 2027.306 for Gregorian year. When we decrease the universe figure to 13,670,000,000, we get the following: 13,670,000,000/360 = 37,972,222.22/2 = 18,986,111.110 /1000 = 18,986.111/3.0 = 6328.704/3.14 = 2015.511 = 2016.0 for Gregorian year—the present year of 2016. Today's biblical year is 6329 and the universe is 13,670,000,000.

In order to determine which form of pi we should use to decode the biblical year 6329 and calculate the age of the universe, we calculate as follows:

a- 6329 x 360 x 2 x 1,000 = 4,556,880,000 x 3.0 = ***13,670,640,000 billion, age of the universe b- 6329 x 360 x 2 x 1,000 = 4,556,880,000 x 3.1 = 14,126,328,000 billion, age of the universe

c- 6329 x 360 x 2 x 1,000 = 4,556,880,000 x 3.14 = 14,308,603,200 billion, age of the universe

We should use (a) 3.0. ***

In order to determine how many years humankind has existed on the earth, we calculate using the following steps: ****

a- 1,000 x 360 (solar year) x 2 x 1,000 = 720,000,000 one creative day

b- 720,000,000 x 7 = 5,040,000,000 billion minus 4,556,880,000 = 483,120,000 million years left in the seventh creative day **** million

c- 720,000,000 minus 483,120,000 = 236,880,000 years

Epilogue and Gratitude

Dear Readers,

We thank you so very much for taking the time to read our book, *New Wine Revelations*. We say we because God has inspired this writer and servant of the Almighty over a blessed period of more than twenty-five years with intuitive thoughts and ideas. We thank God for allowing us to serve the Holy Eloheem, Father, Son, and Holy Spirit. In addition, we greatly thank the countless researchers whom we were able to draw from and confirm our literary work. We pray and hope that none of the contents of our book has offended you in any way. If it has, please forgive us. And feel free to refute any part of NWR.

There are so many more items we would like to put in this book. We won't at this time. Perhaps, and God willing, we will include them in a sequel to this book.

We close by remembering a bedtime prayer taught to us by our mothers and grandmothers: "Now I lay me down to sleep, I pray to God my soul to keep. If I should die before I wake, I pray to God my soul to take." AMEN.

Addenda

Now, in order to solve the dilemma of refuting and rejecting this book after reading it, let us paraphrase a line from Hamlet's soliloquy, "To be or not to be. . ." To refute or not to refute, that is the question. Whether 'tis wiser to refute and reject all new ideas and revelations or to embrace them and investigate them for their undeniable truth. And when we find these truths whether to ignore them or apply them to the countless dilemmas and problems that have deprived us of world peace, harmony among races and nations, confused and plagued generations for centuries? The ball is in our court. Let us *act on it*.

Thank you, Sir William Shakespeare and your theatrical company. May your literary works and dramas be remembered and replayed forever on earth and in Heaven. God bless you forevermore. Amen.

Worldwide fans of the BBC *Dr. Who* drama should be delighted to know that the T.A.R.D.I.S. is an acronym which means Time and Relative Distance in Space.

. . . . . .

Bibliography

Carpenter, Philip L. Professor of Bacteriology. *Chromosomes*. University Of Rhode Island and W. B. Saunders Company (1967): 80, 82, 280, 299.

Carpenter, Philip L. Professor of Bacteriology. *Malaria*. University Of Rhode Island and W. B. Saunders Company (1967): 137, 141.

Carpenter, Philip L. Professor of Bacteriology. *Redox*. University Of Rhode Island and W. B. Saunders Company (1967): 204, 212, 254.

Funk and Wagnalls. New Illustrated Webster's Dictionary/Thesaurus, 1–1150; TH: 1-TH.

Keenan, Charles W. Professor of Chemistry. *Oxidation Potentials*. The University of Tennessee and Jesse H. Wood, Professor Emeritus of Chemistry, the University of Tennessee, General College Chemistry, Harper and Row Publishers, New York (1957): 407, 698–700.

Keenan, Charles W. Professor of Chemistry. *Redox Equations*. The University of Tennessee and Jesse H. Wood, Professor Emeritus of Chemistry, the University of Tennessee, General College Chemistry, Harper and Row Publishers, New York (1957): 423.

Keeton, William T. "Biological Science." *Chromosomes*. Cornell University (1967): 72–78, 500–554.

Keeton, William T. "Biological Science." *Meiosis*. Cornell University (1967): 502–507, 503, 504, 508, 509, 501–509.

Keeton, William T. "Biological Science." *Mitosis.* Cornell University (1967): 502–509.

Solomon, T. W. Graham. "Organic Chemistry." *Ethanol Made From Ethene and Water.* University of South Florida and John Wiley & Sons (1976): 276, 649.

Strong, James. "The New Strong's Exhaustive Concordance of the Bible." *Main Concordance, Hebrew and Chaldee Dictionary.* Universal Subject Guide to Bible, 1990.

The World Book Encyclopedia (Volume C). *Colors.* Field Enterprise Inc. Chicago (First 1938, 1954).

Thomas Nelson, Inc., Holy Bible, New King James Version (NKJV) (1994)

Old Testament (OT)

Genesis, 50 Chapters	Pages 1–78
Exodus, 40 Chapters	Pages 79–142
Leviticus, 27 Chapters	Pages 143–189
Numbers, 36 Chapters	Pages 190–256
Deuteronomy, 34 Chapters	Pages 257–312
Joshua, 24 Chapters	Pages 313–350
2 Chronicles, 36 Chapters	Pages 624–676
Psalm, 150 Chapters	Pages 22:1–22:31
Isaiah, 66 Chapters	Pages 975–1073
Ezekiel, 48 Chapters	Pages 1183–1262 *(4:6)
Daniel, 12 Chapters	Pages 1263–1287
Zephaniah, 14 Chapters	Page 1:15, Darkness in Day
Zechariah, 14 Chapters	Page 14:7, Light in Evening

New Testament (NT)

Matthew, 28 Chapters	Pages 1375–1424
Mark, 16 Chapters	Pages 1425–1456
Luke, 24 Chapters	Pages 1457–1509
John, 21 Chapters	Pages 1510–1548
Acts, 28 Chapters	Pages 1549–1599
James, 5 Chapters	Pages 1715–1719
1 and 2 Peter, 8 Chapters	Pages 1720–1729
1, 2, and 3 John, 7 Chapters	Pages 1730–1737
Jude, 1 Chapter	Pages 1738–1739
Revelations, 22 Chapters	Pages 1740–1763

Other References (OR)

Omniglot/Simon Ager, (1998-2016), The online Encyclopedia of writing systems & languages www.youtube.com/user/omniglot

Wikipedia/Anopheles & Culex malarial mosquitoes

Wikipedia/Book of Revelation Prophecy

Wikipedia/Book of Revelation/1000 years/The millennium

Wikipedia/Cholesterolemia/HDL and LDL increased by the consumption of excessive amounts of shrimp and other non-scaled aquatic creatures; LDL Protein called PCSK-9/Statin drugs treatment

Wikipedia/Colors of the rainbow, list of colors

Wikipedia/Ebola Virus (EVD)/Fruit Bats or Vampire bats carriers in nature (1976) First known case/ decreased blood clotting/ Vitamin K possible treatment among other drugs and procedures

Wikipedia/https:en.wikipedia.org/wiki, The Free Encyclopedia/Age of the Universe, Modified May 05, 2016; Age of the Earth/4.5 +/- 0.05 billion years

Wikipedia/Human skin colors/Color terminology for Race

Wikipedia/Malaria/Plasmodium Falciparum/Female malarial mosquito injects anti-coagulant into host's blood stream (Vitamin K possible treatment)

Wikipedia/Methemoglobinemia/Oxygen carrying Metalloprotein hemoglobin/ Oxidized Ferrous + 2 Iron to Ferric + 3 iron state

Wikipedia/The lost years of Jesus's possible travels to India, Tibet & Babylon

Wikipedia/The Shiva, Judaism mourning/Present practice 7 days plus 30 days; Old practice in Jesus's first Advent, 3 days + 7 days + 30 days = 40 days

Thank you in different languages

Thank you	English
Kamsahamida	Korean
Kia ora	Maori
Obrigado	Portuguese
Spasiba	Russian
Dhanyawadah	Sanskrit
Tapadh leat	Scottish Gaelic
Gracias	Spanish
Asante	Swahili
Tack	Swedish
Tujay chay	Tibetan
Grazie	Italian
Domo arigato	Japanese
Diakuju	Ukrainian
Chokian	Arabic
Osiyo	Cherokee
Salamat	Filipino
Danke	German
Merci	French
Ethansto poli	Greek
Todah rabah	Hebrew

About the Author

My name is George Waymond "Wayne" Chivers II. My father's middle name was Washington. I was born on January 01, 1939, at 3:00 PM in Carrollton, Georgia, fifty miles Southwest of Atlanta, Georgia.

My mother, Laura Gibson Boone, her mother, Fanny Mae Merdis and my father's mother, Mary, were born with the gifts of ESP, aka as clairvoyancy, remote viewing, remote hearing, aka clairaudiency. Often, under a peaceful, prayerful, happy alpha wave state of mind and soul, these ESP gifts and talents also manifest themselves in me. We all have these gifts to a certain degree. They are most effective when we love God and each other.

My genealogy, I'm told, consist of Western and Central African, Welsh, Scott-Irish, English, Seminole, French (Chevre), Cherokee and Italian (Capas).

My interests are Christ unification of all religions, research, health, environment, education and worldwide welfare.

My hobbies are Sudoku, Peanocle, clean jokes, ballroom and Latin dancing. Also, R&D medical and Engineering Technology.

www.ingramcontent.com/pod-product-compliance
Lightning Source LLC
Chambersburg PA
CBHW030913180526
45163CB00004B/1812